Java 数据分析(影印版)
Java Data Analysis

John R. Hubbard 著

南京　东南大学出版社

图书在版编目(CIP)数据

Java 数据分析：英文/(美)约翰·R.哈伯德(John R. Hubbard)著. —影印本. —南京：东南大学出版社，2018.8
书名原文：Java Data Analysis
ISBN 978-7-5641-7736-2

Ⅰ.①J… Ⅱ.①约… Ⅲ.①JAVA语言—程序设计—英文 Ⅳ.①TP312.8

中国版本图书馆 CIP 数据核字(2018)第 099958 号
图字：10-2018-096 号

© 2017 by PACKT Publishing Ltd.

Reprint of the English Edition, jointly published by PACKT Publishing Ltd and Southeast University Press, 2018. Authorized reprint of the original English edition, 2018 PACKT Publishing Ltd, the owner of all rights to publish and sell the same.

All rights reserved including the rights of reproduction in whole or in part in any form.

英文原版由 PACKT Publishing Ltd 出版 2017。

英文影印版由东南大学出版社出版 2018。此影印版的出版和销售得到出版权和销售权的所有者——PACKT Publishing Ltd 的许可。

版权所有，未得书面许可，本书的任何部分和全部不得以任何形式重制。

Java 数据分析(影印版)

出版发行：东南大学出版社
地　　址：南京四牌楼 2 号　邮编：210096
出 版 人：江建中
网　　址：http://www.seupress.com
电子邮件：press@seupress.com
印　　刷：常州市武进第三印刷有限公司
开　　本：787 毫米×980 毫米　16 开本
印　　张：25.75
字　　数：504 千字
版　　次：2018 年 8 月第 1 版
印　　次：2018 年 8 月第 1 次印刷
书　　号：ISBN 978-7-5641-7736-2
定　　价：94.00 元

本社图书若有印装质量问题，请直接与营销部联系。电话(传真)：025-83791830

Credits

Author
John R. Hubbard

Reviewers
Erin Paciorkowski
Alexey Zinoviev

Commissioning Editor
Amey Varangaonkar

Acquisition Editor
Varsha Shetty

Content Development Editor
Aishwarya Pandere

Technical Editor
Prasad Ramesh

Copy Editor
Safis Editing

Project Coordinator
Nidhi Joshi

Proofreader
Safis Editing

Indexer
Tejal Daruwale Soni

Graphics
Tania Dutta

Production Coordinator
Arvindkumar Gupta

Cover Work
Arvindkumar Gupta

About the Author

John R. Hubbard has been doing computer-based data analysis for over 40 years at colleges and universities in Pennsylvania and Virginia. He holds an MSc in computer science from Penn State University and a PhD in mathematics from the University of Michigan. He is currently a professor of mathematics and computer science, Emeritus, at the University of Richmond, where he has been teaching data structures, database systems, numerical analysis, and big data.

Dr. Hubbard has published many books and research papers, including six other books on computing. Some of these books have been translated into German, French, Chinese, and five other languages. He is also an amateur timpanist.

> I would like to thank the reviewers of this book for their valuable comments and suggestions. I would also like to thank the energetic team at Packt for publishing the book and helping me perfect it. Finally, I would like to thank my family for supporting me through everything.

About the Reviewers

Erin Paciorkowski studied computer science at the Georgia Institute of Technology as a National Merit Scholar. She has worked in Java development for the Department of Defense for over 8 years and is also a graduate teaching assistant for the Georgia Tech Online Masters of Computer Science program. She is a certified scrum master and holds Security+, Project+, and ITIL Foundation certifications. She was a Grace Hopper Celebration Scholar in 2016. Her interests include data analysis and information security.

Alexey Zinoviev is a lead engineer and Java and big data trainer at EPAM Systems, with a focus on Apache Spark, Apache Kafka, Java concurrency, and JVM internals. He has deep expertise in machine learning, large graph processing, and the development of distributed scalable Java applications. You can follow him at `@zaleslaw` or `https://github.com/zaleslaw`.

Currently, he's working on a Spark Tutorial at `https://github.com/zaleslaw/Spark-Tutorial` and on an Open GitBook about Spark (in Russian) at `https://zaleslaw.gitbooks.io/data-processing-book/content/`.

> Thanks to my wife, Anastasya, and my little son, Roman, for quietly tolerating the very long hours I've been putting into this book.

www.PacktPub.com

eBooks, discount offers, and more

Did you know that Packt offers eBook versions of every book published, with PDF and ePub files available? You can upgrade to the eBook version at `www.PacktPub.com` and as a print book customer, you are entitled to a discount on the eBook copy. Get in touch with us at `customercare@packtpub.com` for more details.

At `www.PacktPub.com`, you can also read a collection of free technical articles, sign up for a range of free newsletters and receive exclusive discounts and offers on Packt books and eBooks.

https://www.packtpub.com/mapt

Get the most in-demand software skills with Mapt. Mapt gives you full access to all Packt books and video courses, as well as industry-leading tools to help you plan your personal development and advance your career.

Why subscribe?

- Fully searchable across every book published by Packt
- Copy and paste, print, and bookmark content
- On demand and accessible via a web browser

Customer Feedback

Thanks for purchasing this Packt book. At Packt, quality is at the heart of our editorial process. To help us improve, please leave us an honest review on this book's Amazon page at `https://www.amazon.com/dp/1787285650`.

If you'd like to join our team of regular reviewers, you can e-mail us at `customerreviews@packtpub.com`. We award our regular reviewers with free eBooks and videos in exchange for their valuable feedback. Help us be relentless in improving our products!

Table of Contents

Preface	**vii**
Chapter 1: Introduction to Data Analysis	**1**
Origins of data analysis	1
The scientific method	2
Actuarial science	3
Calculated by steam	3
A spectacular example	4
Herman Hollerith	5
ENIAC	6
VisiCalc	7
Data, information, and knowledge	8
Why Java?	8
Java Integrated Development Environments	9
Summary	11
Chapter 2: Data Preprocessing	**13**
Data types	13
Variables	14
Data points and datasets	14
Null values	15
Relational database tables	15
Key fields	15
Key-value pairs	16
Hash tables	16
File formats	18
Microsoft Excel data	20
XML and JSON data	24
Generating test datasets	30
Metadata	31

    Data cleaning   32
    Data scaling   32
    Data filtering   33
    Sorting   36
    Merging   38
    Hashing   41
  Summary   **42**

Chapter 3: Data Visualization   45

  Tables and graphs   **46**
    Scatter plots   46
    Line graphs   48
    Bar charts   49
    Histograms   50
  Time series   **51**
  Java implementation   **53**
  Moving average   **56**
  Data ranking   **61**
  Frequency distributions   **63**
  The normal distribution   **65**
    A thought experiment   66
  The exponential distribution   **68**
  Java example   **69**
  Summary   **71**

Chapter 4: Statistics   73

  Descriptive statistics   **73**
  Random sampling   **76**
  Random variables   **79**
  Probability distributions   **79**
  Cumulative distributions   **81**
  The binomial distribution   **83**
  Multivariate distributions   **87**
  Conditional probability   **89**
  The independence of probabilistic events   **90**
  Contingency tables   **91**
  Bayes' theorem   **91**
  Covariance and correlation   **93**
  The standard normal distribution   **96**
  The central limit theorem   **102**
  Confidence intervals   **103**

Hypothesis testing	**105**
Summary	**107** |//

Chapter 5: Relational Databases — 109

The relation data model	**109**
Relational databases	**110**
Foreign keys	**111**
Relational database design	**112**
Creating a database	114
SQL commands	117
Inserting data into the database	121
Database queries	123
SQL data types	125
JDBC	125
Using a JDBC PreparedStatement	128
Batch processing	130
Database views	134
Subqueries	137
Table indexes	140
Summary	**142**

Chapter 6: Regression Analysis — 143

Linear regression	**143**
Linear regression in Excel	144
Computing the regression coefficients	148
Variation statistics	151
Java implementation of linear regression	155
Anscombe's quartet	163
Polynomial regression	**165**
Multiple linear regression	171
The Apache Commons implementation	174
Curve fitting	176
Summary	**178**

Chapter 7: Classification Analysis — 179

Decision trees	**180**
What does entropy have to do with it?	181
The ID3 algorithm	185
Java Implementation of the ID3 algorithm	195
The Weka platform	198
The ARFF filetype for data	198
Java implementation with Weka	201

Bayesian classifiers	**203**
Java implementation with Weka	206
Support vector machine algorithms	209
Logistic regression	**214**
K-Nearest Neighbors	220
Fuzzy classification algorithms	225
Summary	**225**
Chapter 8: Cluster Analysis	**227**
Measuring distances	**227**
The curse of dimensionality	**233**
Hierarchical clustering	**235**
Weka implementation	245
K-means clustering	248
K-medoids clustering	253
Affinity propagation clustering	256
Summary	**265**
Chapter 9: Recommender Systems	**267**
Utility matrices	**268**
Similarity measures	**270**
Cosine similarity	**271**
A simple recommender system	**272**
Amazon's item-to-item collaborative filtering recommender	**284**
Implementing user ratings	**290**
Large sparse matrices	**294**
Using random access files	**298**
The Netflix prize	**300**
Summary	**301**
Chapter 10: NoSQL Databases	**303**
The Map data structure	**303**
SQL versus NoSQL	**306**
The Mongo database system	**307**
The Library database	**314**
Java development with MongoDB	**318**
The MongoDB extension for geospatial databases	**327**
Indexing in MongoDB	**328**
Why NoSQL and why MongoDB?	**329**
Other NoSQL database systems	**329**
Summary	**330**

Chapter 11: Big Data Analysis with Java — 331
- Scaling, data striping, and sharding — 331
- Google's PageRank algorithm — 332
- Google's MapReduce framework — 338
- Some examples of MapReduce applications — 339
- The WordCount example — 339
- Scalability — 344
- Matrix multiplication with MapReduce — 346
- MapReduce in MongoDB — 350
- Apache Hadoop — 352
- Hadoop MapReduce — 353
- Summary — 354

Appendix: Java Tools — 355
- The command line — 355
- Java — 358
- NetBeans — 360
- MySQL — 364
- MySQL Workbench — 365
- Accessing the MySQL database from NetBeans — 373
- The Apache Commons Math Library — 376
- The javax JSON Library — 380
- The Weka libraries — 381
- MongoDB — 382

Index — 385

Preface

"It has been said that you don't really understand something until you have taught it to someone else. The truth is that you don't really understand it until you have taught it to a computer; that is, implemented it as an algorithm."

— *Donald Knuth*

As Don Knuth so wisely said, the best way to understand something is to implement it. This book will help you understand some of the most important algorithms in data science by showing you how to implement them in the Java programming language.

The algorithms and data management techniques presented here are often categorized under the general fields of data science, data analytics, predictive analytics, artificial intelligence, business intelligence, knowledge discovery, machine learning, data mining, and big data. We have included many that are relatively new, surprisingly powerful, and quite exciting. For example, the ID3 classification algorithm, the K-means and K-medoid clustering algorithms, Amazon's recommender system, and Google's PageRank algorithm have become ubiquitous in their effect on nearly everyone who uses electronic devices on the web.

We chose the Java programming language because it is the most widely used language and because of the reasons that make it so: it is available, free, everywhere; it is object-oriented; it has excellent support systems, such as powerful integrated development environments; its documentation system is efficient and very easy to use; and there is a multitude of open source libraries from third parties that support essentially all implementations that a data analyst is likely to use. It's no coincidence that systems such as MongoDB, which we study in *Chapter 11, Big Data Analysis with Java*, are themselves written in Java.

What this book covers

Chapter 1, Introduction to Data Analysis, introduces the subject, citing its historical development and its importance in solving critical problems of the society.

Chapter 2, Data Preprocessing, describes the various formats for data storage, the management of datasets, and basic preprocessing techniques such as sorting, merging, and hashing.

Chapter 3, Data Visualization, covers graphs, charts, time series, moving averages, normal and exponential distributions, and applications in Java.

Chapter 4, Statistics, reviews fundamental probability and statistical principles, including randomness, multivariate distributions, binomial distribution, conditional probability, independence, contingency tables, Bayes' theorem, covariance and correlation, central limit theorem, confidence intervals, and hypothesis testing.

Chapter 5, Relational Databases, covers the development and access of relational databases, including foreign keys, SQL, queries, JDBC, batch processing, database views, subqueries, and indexing. You will learn how to use Java and JDBC to analyze data stored in relational databases.

Chapter 6, Regression Analysis, demonstrates an important part of predictive analysis, including linear, polynomial, and multiple linear regression. You will learn how to implement these techniques in Java using the Apache Commons Math library.

Chapter 7, Classification Analysis, covers decision trees, entropy, the ID3 algorithm and its Java implementation, ARFF files, Bayesian classifiers and their Java implementation, support vector machine (SVM) algorithms, logistic regression, K-nearest neighbors, and fuzzy classification algorithms. You will learn how to implement these algorithms in Java with the Weka library.

Chapter 8, Cluster Analysis, includes hierarchical clustering, K-means clustering, K-medoids clustering, and affinity propagation clustering. You will learn how to implement these algorithms in Java with the Weka library.

Chapter 9, Recommender Systems, covers utility matrices, similarity measures, cosine similarity, Amazon's item-to-item recommender system, large sparse matrices, and the historic Netflix Prize competition.

Chapter 10, NoSQL Databases, centers on the MongoDB database system. It also includes geospatial databases and Java development with MongoDB.

Chapter 11, *Big Data Analysis*, covers Google's PageRank algorithm and its MapReduce framework. Particular attention is given to the complete Java implementations of two characteristic examples of MapReduce: WordCount and matrix multiplication.

Appendix, *Java Tools*, walks you through the installation of all of the software used in the book: NetBeans, MySQL, Apache Commons Math Library, javax.json, Weka, and MongoDB.

What you need for this book

This book is focused on an understanding of the fundamental principles and algorithms used in data analysis. This understanding is developed through the implementation of those principles and algorithms in the Java programming language. Accordingly, the reader should have some experience of programming in Java. Some knowledge of elementary statistics and some experience with database work will also be helpful.

Who this book is for

This book is for both students and practitioners who seek to further their understanding of data analysis and their ability to develop Java software that implements algorithms in that field.

Conventions

In this book, you will find a number of text styles that distinguish between different kinds of information. Here are some examples of these styles and an explanation of their meaning.

Code words in text, database table names, folder names, filenames, file extensions, pathnames, dummy URLs, user input, and Twitter handles are shown as follows: "We can include other contexts through the use of the `include` directive."

A block of code is set as follows:

```
Color = {RED, YELLOW, BLUE, GREEN, BROWN, ORANGE}
Surface = {SMOOTH, ROUGH, FUZZY}
Size = {SMALL, MEDIUM, LARGE}
```

Any command-line input or output is written as follows:

```
mongo-java-driver-3.4.2.jar
mongo-java-driver-3.4.2-javadoc.jar
```

New terms and **important words** are shown in bold. Words that you see on the screen, for example, in menus or dialog boxes, appear in the text like this: "Clicking the **Next** button moves you to the next screen."

> Warnings or important notes appear in a box like this.

> Tips and tricks appear like this.

Reader feedback

Feedback from our readers is always welcome. Let us know what you think about this book—what you liked or disliked. Reader feedback is important for us as it helps us develop titles that you will really get the most out of.

To send us general feedback, simply e-mail feedback@packtpub.com, and mention the book's title in the subject of your message.

If there is a topic that you have expertise in and you are interested in either writing or contributing to a book, see our author guide at www.packtpub.com/authors.

Customer support

Now that you are the proud owner of a Packt book, we have a number of things to help you to get the most from your purchase.

Downloading the example code

You can download the example code files for this book from your account at http://www.packtpub.com. If you purchased this book elsewhere, you can visit http://www.packtpub.com/support and register to have the files e-mailed directly to you.

You can download the code files by following these steps:

1. Log in or register to our website using your e-mail address and password.
2. Hover the mouse pointer on the **SUPPORT** tab at the top.
3. Click on **Code Downloads & Errata**.
4. Enter the name of the book in the **Search** box.
5. Select the book for which you're looking to download the code files.
6. Choose from the drop-down menu where you purchased this book from.
7. Click on **Code Download**.

You can also download the code files by clicking on the **Code Files** button on the book's webpage at the Packt Publishing website. This page can be accessed by entering the book's name in the **Search** box. Please note that you need to be logged in to your Packt account.

Once the file is downloaded, please make sure that you unzip or extract the folder using the latest version of:

- WinRAR / 7-Zip for Windows
- Zipeg / iZip / UnRarX for Mac
- 7-Zip / PeaZip for Linux

The code bundle for the book is also hosted on GitHub at https://github.com/PacktPublishing/Java-Data-Analysis. We also have other code bundles from our rich catalog of books and videos available at https://github.com/PacktPublishing/. Check them out!

Errata

Although we have taken every care to ensure the accuracy of our content, mistakes do happen. If you find a mistake in one of our books—maybe a mistake in the text or the code—we would be grateful if you could report this to us. By doing so, you can save other readers from frustration and help us improve subsequent versions of this book. If you find any errata, please report them by visiting http://www.packtpub.com/submit-errata, selecting your book, clicking on the **Errata Submission Form** link, and entering the details of your errata. Once your errata are verified, your submission will be accepted and the errata will be uploaded to our website or added to any list of existing errata under the Errata section of that title.

To view the previously submitted errata, go to https://www.packtpub.com/books/content/support and enter the name of the book in the search field. The required information will appear under the **Errata** section.

Piracy

Piracy of copyrighted material on the Internet is an ongoing problem across all media. At Packt, we take the protection of our copyright and licenses very seriously. If you come across any illegal copies of our works in any form on the Internet, please provide us with the location address or website name immediately so that we can pursue a remedy.

Please contact us at `copyright@packtpub.com` with a link to the suspected pirated material.

We appreciate your help in protecting our authors and our ability to bring you valuable content.

Questions

If you have a problem with any aspect of this book, you can contact us at `questions@packtpub.com`, and we will do our best to address the problem.

1
Introduction to Data Analysis

Data analysis is the process of organizing, cleaning, transforming, and modeling data to obtain useful information and ultimately, new knowledge. The terms data analytics, business analytics, data mining, artificial intelligence, machine learning, knowledge discovery, and big data are also used to describe similar processes. The distinctions of these fields probably lie more in their areas of application than in their fundamental nature. Some argue that these are all part of the new discipline of data science.

The central process of gaining useful information from organized data is managed by the application of computer science algorithms. Consequently, these will be a central focus of this book.

Data analysis is both an old field and a new one. Its origins lie among the mathematical fields of numerical methods and statistical analysis, which reach back into the eighteenth century. But many of the methods that we shall study gained prominence much more recently, with the ubiquitous force of the internet and the consequent availability of massive datasets.

In this first chapter, we look at a few famous historical examples of data analysis. These can help us appreciate the importance of the science and its promise for the future.

Origins of data analysis

Data is as old as civilization itself, maybe even older. The 17,000-year-old paintings in the Lascaux caves in France could well have been attempts by those primitive dwellers to record their greatest hunting triumphs. Those records provide us with data about humanity in the Paleolithic era. That data was not analyzed, in the modern sense, to obtain new knowledge. But its existence does attest to the need humans have to preserve their ideas in data.

Five thousand years ago, the Sumerians of ancient Mesopotamia recorded far more important data on clay tablets. That cuneiform writing included substantial accounting data about daily business transactions. To apply that data, the Sumerians invented not only text writing, but also the first number system.

In 1086, King William the Conqueror ordered a massive collection of data to determine the extent of the lands and properties of the crown and of his subjects. This was called the *Domesday Book*, because it was a final tallying of people's (material) lives. That data was analyzed to determine ownership and tax obligations for centuries to follow.

The scientific method

On November 11, 1572, a young Danish nobleman named Tycho Brahe observed the supernova of a star that we now call **SN 1572**. From that time until his death 30 years later, he devoted his wealth and energies to the accumulation of astronomical data. His young German assistant, Johannes Kepler, spent 18 years analyzing that data before he finally formulated his three laws of planetary motion in 1618.

Figure 1 Kepler

Historians of science usually attribute Kepler's achievement as the beginning of the Scientific Revolution. Here were the essential steps of the scientific method: observe nature, collect the data, analyze the data, formulate a theory, and then test that theory with more data. Note the central step here: data analysis.

Of course, Kepler did not have either of the modern tools that data analysts use today: algorithms and computers on which to implement them. He did, however, apply one technological breakthrough that surely facilitated his number crunching: logarithms. In 1620, he stated that Napier's invention of logarithms in 1614 had been essential to his discovery of the third law of planetary motion.

Kepler's achievements had a profound effect upon Galileo Galilei a generation later, and upon Isaac Newton a generation after him. Both men practiced the scientific method with spectacular success.

Actuarial science

One of Newton's few friends was Edmund Halley, the man who first computed the orbit of his eponymous comet. Halley was a polymath, with expertise in astronomy, mathematics, physics, meteorology, geophysics, and cartography.

In 1693, Halley analyzed mortality data that had been compiled by Caspar Neumann in Breslau, Germany. Like Kepler's work with Brahe's data 90 years earlier, Halley's analysis led to new knowledge. His published results allowed the British government to sell life annuities at the appropriate price, based on the age of the annuitant.

Most data today is still numeric. But most of the algorithms we will be studying apply to a much broader range of possible values, including text, images, audio and video files, and even complete web pages on the internet.

Calculated by steam

In 1821, a young Cambridge student named Charles Babbage was poring over some trigonometric and logarithmic tables that had been recently computed by hand. When he realized how many errors they had, he exclaimed, "*I wish to God these calculations had been executed by steam.*" He was suggesting that the tables could have been computed automatically by some mechanism that would be powered by a steam engine.

Babbage was a mathematician by avocation, holding the same Lucasian Chair of Mathematics at Cambridge University that Isaac Newton had held 150 years earlier and that Stephen Hawking would hold 150 years later. However, he spent a large part of his life working on automatic computing. Having invented the idea of a programmable computer, he is generally regarded as the first computer scientist. His assistant, Lady Ada Lovelace, has been recognized as the first computer programmer.

Babbage's goal was to build a machine that could analyze data to obtain useful information, the central step of data analysis. By automating that step, it could be carried out on much larger datasets and much more rapidly. His interest in trigonometric and logarithmic tables was related to his objective of improving methods of navigation, which was critical to the expanding British Empire.

A spectacular example

In 1854, cholera broke out among the poor in London. The epidemic spread quickly, partly because nobody knew the source of the problem. But a physician named John Snow suspected it was caused by contaminated water. At that time, most Londoners drew their water from public wells that were supplied directly from the River Thames. The following figure shows the map that Snow drew, with black rectangles indicating the frequencies of cholera occurrences:

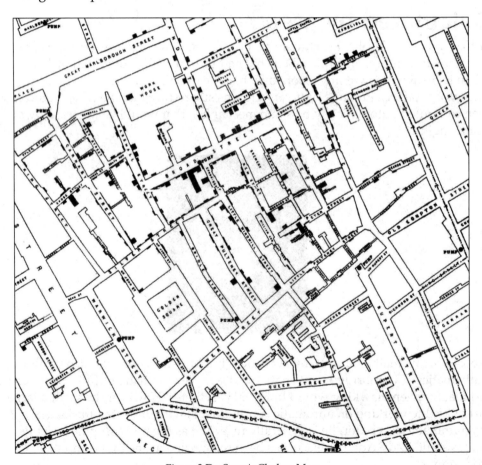

Figure 3 Dr. Snow's Cholera Map

If you look closely, you can also see the locations of nine public water pumps, marked as black dots and labeled **PUMP**. From this data, we can easily see that the pump at the corner of Broad Street and Cambridge Street is in the middle of the epidemic. This data analysis led Snow to investigate the water supply at that pump, discovering that raw sewage was leaking into it through a break in the pipe.

By also locating the public pumps on the map, he demonstrated that the source was probably the pump at the corner of Broad Street and Cambridge Street. This was one of the first great examples of the successful application of data analysis to public health (for more information, see `https://www1.udel.edu/johnmack/frec682/cholera/cholera2.html`). President James K. Polk and composer Pyotr Ilyich Tchaikovsky were among the millions who died from cholera in the nineteenth century. But even today the disease is still a pandemic, killing around 100,000 per year world-wide.

Herman Hollerith

The decennial United States Census was mandated by the U. S. Constitution in 1789 for the purposes of apportioning representatives and taxes. The first census was taken in 1790 when the U. S. population was under four million. It simply counted free men. But by 1880, the country had grown to over 50 million, and the census itself had become much more complicated, recording dependents, parents, places of birth, property, and income.

Figure 4 Hollerith

The 1880 census took over eight years to compile. The United States Census Bureau realized that some sort of automation would be required to complete the 1890 census. They hired a young engineer named Herman Hollerith, who had proposed a system of electronic tabulating machines that would use punched cards to record the data.

This was the first successful application of automated data processing. It was a huge success. The total population of nearly 62 million was reported after only six weeks of tabulation.

Hollerith was awarded a Ph.D. from MIT for his achievement. In 1911, he founded the Computing-Tabulating-Recording Company, which became the **International Business Machines Corporation (IBM)** in 1924. Recently IBM built the supercomputer Watson, which was probably the most successful commercial application of data mining and artificial intelligence yet produced.

ENIAC

During World War II, the U. S. Navy had battleships with guns that could shoot 2700-pound projectiles 24 miles. At that range, a projectile spent almost 90 seconds in flight. In addition to the guns' elevation, angle of amplitude, and initial speed of propulsion, those trajectories were also affected by the motion of the ship, the weather conditions, and even the motion of the earth's rotation. Accurate calculations of those trajectories posed great problems.

To solve these computational problems, the U. S. Army contracted an engineering team at the University of Pennsylvania to build the **Electronic Numerical Integrator and Computer (ENIAC)**, the first complete electronic programmable digital computer. Although not completed until after the war was over, it was a huge success.

It was also enormous, occupying a large room and requiring a staff of engineers and programmers to operate. The input and output data for the computer were recorded on Hollerith cards. These could be read automatically by other machines that could then print their contents.

ENIAC played an important role in the development of the hydrogen bomb. Instead of artillery tables, it was used to simulate the first test run for the project. That involved over a million cards.

Chapter 1

Figure 5 ENIAC

VisiCalc

In 1979, Harvard student Dan Bricklin was watching his professor correct entries in a table of finance data on a chalkboard. After correcting a mistake in one entry, the professor proceeded to correct the corresponding marginal entries. Bricklin realized that such tedious work could be done much more easily and accurately on his new Apple II microcomputer. This resulted in his invention of VisiCalc, the first spreadsheet computer program for microcomputers. Many agree that that innovation transformed the microcomputer from a hobbyist's game platform to a serious business tool.

The consequence of Bricklin's VisiCalc was a paradigm shift in commercial computing. Spreadsheet calculations, an essential form of commercial data processing, had until then required very large and expensive mainframe computing centers. Now they could be done by a single person on a personal computer. When the IBM PC was released two years later, VisiCalc was regarded as essential software for business and accounting.

Data, information, and knowledge

The 1854 cholera epidemic case is a good example for understanding the differences between data, information, and knowledge. The data that Dr. Snow used, the locations of cholera outbreaks and water pumps, was already available. But the connection between them had not yet been discovered. By plotting both datasets on the same city map, he was able to determine that the pump at Broad street and Cambridge street was the source of the contamination. That connection was new information. That finally led to the new knowledge that the disease is transmitted by foul water, and thus the new knowledge on how to prevent the disease.

Why Java?

Java is, as it has been for over a decade, the most popular programming language in the world. And its popularity is growing. There are several good reasons for this:

- Java runs the same way on all computers
- It supports the object-oriented programming (OOP) paradigm
- It interfaces easily with other languages, including the database query language SQL
- Its Javadoc documentation is easy to access and use
- Most open-source software is written in Java, including that which is used for data analysis

Python may be easier to learn, R may be simpler to run, JavaScript may be easier for developing websites, and C/C++ may be faster, but for general purpose programming, Java can't be beat.

Java was developed in 1995 by a team led by James Gosling at Sun Microsystems. In 2010, the Oracle Corporation bought Sun for $7.4 B and has supported Java since then. The current version is Java 8, released in 2014. But by the time you buy this book, Java 9 should be available; it is scheduled to be released in late 2017.

As the title of this book suggests, we will be using Java in all our examples.

[*Appendix* includes instructions on how to set up your computer with Java.]

Java Integrated Development Environments

To simplify Java software development, many programmers use an **Integrated Development Environment** (**IDE**). There are several good, free Java IDEs available for download. Among them are:

- NetBeans
- Eclipse
- JDeveloper
- JCreator
- IntelliJ IDEA

These are quite similar in how they work, so once you have used one, it's easy to switch to another.

Although all the Java examples in this book can be run at the command line, we will instead show them running on NetBeans. This has several advantages, including:

- Code listings include line numbers
- Standard indentation rules are followed automatically
- Code syntax coloring

Here is the standard Hello World program in NetBeans:

```java
/*  Data Analysis with Java
 *  John R. Hubbard
 *  March 30, 2017
 */

package dawj.ch01;

public class HelloWorld {
    public static void main(String[] args) {
        System.out.println("Hello, World!");
    }
}
```

Listing 1 Hello World program

When you run this program in NetBeans, you will see some of its syntax coloring: gray for comments, blue for reserved words, green for objects, and orange for strings.

In most cases, to save space, we will omit the header comments and the package designation from the listing displays, showing only the program, like this:

```
HelloWorld.java
 8  public class HelloWorld {
 9      public static void main(String[] args) {
10          System.out.println("Hello, World!");
11      }
12  }
```

Listing 2 Hello World program abbreviated

Or, sometimes just we'll show the `main()` method, like this:

```
 9      public static void main(String[] args) {
10          System.out.println("Hello, World!");
11      }
```

Listing 3 Hello World program abbreviated further

Nevertheless, all the complete source code files are available for download at the Packt Publishing website.

Here is the output from the Hello World program:

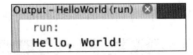

Figure 6 Output from the Hello World program

[*Appendix* describes how to install and start using NetBeans.]

Summary

The first part of this chapter described some important historical events that have led to the development of data analysis: ancient commercial record keeping, royal compilations of land and property, and accurate mathematical models in astronomy, physics, and navigation. It was this activity that led Babbage to invent the computer. Data analysis was borne from necessity in the advance of civilization, from the identification of the source of cholera, to the management of economic data, and the modern processing of massive datasets.

This chapter also briefly explained our choice of the Java programming language for the implementation of the data analysis algorithms to be studied in this book. And finally, it introduced the NetBeans IDE, which we will also use throughout the book.

2
Data Preprocessing

Before data can be analyzed, it is usually processed into some standardized form. This chapter describes those processes.

Data types

Data is categorized into types. A data type identifies not only the form of the data but also what kind of operations can be performed upon it. For example, arithmetic operations can be performed on numerical data, but not on text data.

A data type can also determine how much computer storage space an item requires. For example, a decimal value like 3.14 would normally be stored in a 32-bit (four bytes) slot, while a web address such as https://google.com might occupy 160 bits.

Here is a categorization of the main data types that we will be working with in this book. The corresponding Java types are shown in parentheses:

- Numeric types
    - Integer (int)
    - Decimal (double)
- Text type
    - String (String)
- Object types
    - Date (java.util.Date)
    - File (java.io.File)
    - General object (Object)

Variables

In computer science, we think of a variable as a storage location that holds a data value. In Java, a variable is introduced by declaring it to have a specific type. For example, consider the following statement:

```
String lastName;
```

It declares the variable `lastName` to have type `String`.

We can also initialize a variable with an explicit value when it is declared, like this:

```
double temperature = 98.6;
```

Here, we would think of a storage location named `temperature` that contains the value `98.6` and has type `double`.

Structured variables can also be declared and initialized in the same statement:

```
int[] a = {88, 11, 44, 77, 22};
```

This declares the variable `a` to have type `int[]` (array of `int`s) and contain the five elements specified.

Data points and datasets

In data analysis, it is convenient to think of the data as points of information. For example, in a collection of biographical data, each data point would contain information about one person. Consider the following data point:

```
("Adams", "John", "M", 26, 704601929)
```

It could represent a 26-year-old male named `John Adams` with ID number `704601929`.

We call the individual data values in a data point **fields** (or **attributes**). Each of these values has its own type. The preceding example has five fields: three text and two numeric.

The sequence of data types for the fields of a data point is called its **type signature**. The type signature for the preceding example is (text, text, text, numeric, numeric). In Java, that type signature would be (`String`, `String`, `String`, `int`, `int`).

A **dataset** is a set of data points, all of which have the same type signature. For example, we could have a dataset that represents a group of people, each point representing a unique member of the group. Since all points of the set have the same type signature, we say that signature characterizes the dataset itself.

Null values

There is one special data value whose type is unspecified, and therefore may take the role of any type. That is the null value. It usually means unknown. So, for example, the preceding dataset described could contain the data point ("White", null, "F", 39, 440163867), which would represent a 39-year-old female with last name White and ID number 440163867 but whose first name is unknown (or unspecified).

Relational database tables

In a relational database, we think of each dataset as a table, with each data point being a row in the table. The dataset's signature defines the columns of the table.

Here is an example of a relational database table. It has four rows and five columns, representing a dataset of four data points with five fields:

Last name	First name	Sex	Age	ID
Adams	John	M	26	704601929
White	null	F	39	440163867
Jones	Paul	M	49	602588410
Adams	null	F	30	120096334

 There are two null fields in this table.

Because a database table is really a set of rows, the order of the rows is irrelevant, just as the order of the data points in any dataset is irrelevant. For the same reason, a database table may not contain duplicate rows and a dataset may not contain duplicate data points.

Key fields

A dataset may specify that all values of a designated field be unique. Such a field is called a **key field** for the dataset. In the preceding example, the ID number field could serve as a key field.

When specifying a key field for a dataset (or a database table), the designer must anticipate what kind of data the dataset might hold. For example, the *First name* field in the preceding table would be a bad choice for a key field, because many people have the same first name.

Key values are used for searching. For example, if we want to know how old Paul Jones is, we would first find the data point (that is, row) whose *ID* is *602588410* and then check the age value for that point.

A dataset may designate a subset of fields, instead of a single field, as its key. For example, in a dataset of geographical data, we might designate the `Longitude` and `Latitude` fields together as forming the key.

Key-value pairs

A dataset which has specified a subset of fields (or a single field) as the key is often thought of as a set of **key-value pairs** (**KVP**). From this point of view, each data point has two parts: its key and its value for that key. (The term **attribute-value pairs** is also used.)

In the preceding example, the key is *ID* and the value is *Last name, First name, Sex, Age*.

In the previously mentioned geographical dataset, the key could be `Longitude`, `Latitude` and the value could be `Altitude`, `Average Temperature`, `Average`, and `Precipitation`.

We sometimes think of a key-value pairs dataset as an input-output structure, with the key fields as input and the value fields as output. For example, in the geographical dataset, given a longitude and a latitude, the dataset returns the corresponding altitude, average temperature, and average precipitation.

Hash tables

A dataset of key-value pairs is usually implemented as a hash table. It is a data structure in which the key acts like an index into the set, much like page numbers in a book or line numbers in a table. This direct access is much faster than sequential access, which is like searching through a book page-by-page for a certain word or phrase.

In Java, we usually use the `java.util.HashMap<Key,Value>` class to implement a key-value pair dataset. The type parameters `Key` and `Value` are specified classes. (There is also an older `HashTable` class, but it is considered obsolete.)

Here is a data file of seven South American countries:

```
Countries.dat
Argentina     41,343,201
Brazil       201,103,330
Chile         16,746,491
Columbia      47,790,000
Paraguay       6,375,830
Peru          29,907,003
Venezuela     27,223,228
```

Figure 2-1 Countries data file

Here is a Java program that loads this data into a `HashMap` object:

```
Countries.dat   HashMapExample.java
13 public class HashMapExample {
14     public static void main(String[] args) {
15         File dataFile = new File("data/Countries.dat");
16         HashMap<String,Integer> dataset = new HashMap();
17         try {
18             Scanner input = new Scanner(dataFile);
19             while (input.hasNext()) {
20                 String country = input.next();
21                 int population = input.nextInt();
22                 dataset.put(country, population);
23             }
24         } catch (FileNotFoundException e) {
25             System.out.println(e);
26         }
27         System.out.printf("dataset.size(): %d%n", dataset.size());
28         System.out.printf("dataset.get(\"Peru\"): %,d%n", dataset.get("Peru"));
29     }
30 }
```

Listing 2-1 HashMap example for Countries data

The `Countries.dat` file is in the `data` folder. Line 15 instantiates a `java.io.File` object named `dataFile` to represent the file. Line 16 instantiates a `java.util.HashMap` object named `dataset`. It is structured to have `String` type for its key and `Integer` type for its value. Inside the `try` block, we instantiate a `Scanner` object to read the file. Each data point is loaded at line 22, using the `put()` method of the `HashMap` class.

After the dataset has been loaded, we print its size at line 27 and then print the value for `Peru` at line 28. (The format code `%,d` prints the integer value with commas.)

Here is the program's output:

```
Output - HashMapExample (run)
dataset.size(): 7
dataset.get("Peru"): 29,907,003
```

Figure 2-2 Output from HashMap program

Notice, in the preceding example, how the `get()` method implements the input-output nature of the hash table's key-value structure. At line 28, the input is the name `Peru` and the resulting output is its population 29,907,003.

It is easy to change the value of a specified key in a hash table. Here is the same `HashMap` example with these three more lines of code added:

```
29    dataset.put("Peru", 31000000);
30    System.out.printf("dataset.size(): %d%n", dataset.size());
31    System.out.printf("dataset.get(\"Peru\"): %,d%n", dataset.get("Peru"));
```

Listing 2-2 HashMap example revised

Line 29 changes the value for `Peru` to be 31,000,000.

Here's the output from this revised program:

```
Output - HashMapExample (run)
dataset.size(): 7
dataset.get("Peru"): 29,907,003
dataset.size(): 7
dataset.get("Peru"): 31,000,000
```

Figure 2-3 Output from revised HashMap program

Notice that the size of the hash table remains the same. The `put()` method adds a new data point only when the key value is new.

File formats

The `Countries.dat` data file in the preceding example is a flat file—an ordinary text file with no special structure or formatting. It is the simplest kind of data file.

Another simple, common format for data files is the **comma separated values** (CSV) file. It is also a text file, but uses commas instead of blanks to separate the data values. Here is the same data as before, in CSV format:

Figure 2-4 A CSV data file

 In this example, we have added a header line that identifies the columns by name: Country and Population.

For Java to process this correctly, we must tell the Scanner object to use the comma as a delimiter. This is done at line 18, right after the input object is instantiated:

```java
public class ReadingCSVFiles {
    public static void main(String[] args) {
        File dataFile = new File("data/Countries.csv");
        try {
            Scanner input = new Scanner(dataFile);
            input.useDelimiter(",|\\s");
            String column1 = input.next();
            String column2 = input.next();
            System.out.printf("%-10s%12s%n", column1, column2);
            while (input.hasNext()) {
                String country = input.next();
                int population = input.nextInt();
                System.out.printf("%-10s%,12d%n", country, population);
            }
        } catch (FileNotFoundException e) {
            System.out.println(e);
        }
    }
}
```

Listing 2-3 A program for reading CSV data

The regular expression , |\\s means *comma or any white space*. The Java symbol for white space (blanks, tabs, newline, and so on.) is denoted by '\s'. When used in a string, the backslash character itself must be escaped with another preceding backslash, like this: \\s. The pipe character | means "or" in regular expressions.

Here is the output:

```
Output - ReadingCSVFiles (run)
run:
Country     Population
Argentina   41,343,201
Brazil     201,103,330
Chile       16,746,491
Columbia    47,790,000
Paraguay     6,375,830
Peru        29,907,003
Venezuela   27,223,228
```

Figure 2-5 Output from the CSV program

The format code %-10s means to print the string in a 10-column field, left-justified. The format code %,12d means to print the decimal integer in a 12-column field, right-justified, with a comma preceding each triple of digits (for readability).

Microsoft Excel data

The best way to read from and write to Microsoft Excel data files is to use the POI open source API library from the Apache Software Foundation. You can download the library here: https://poi.apache.org/download.html. Choose the current poi-bin zip file.

This section shows two Java programs for copying data back and forth between a Map data structure and an Excel workbook file. Instead of a HashMap, we use a TreeMap, just to show how the latter keeps the data points in order by their key values.

The first program is named FromMapToExcel.java. Here is its main() method:

```
Countries.dat   FromMapToExcel.java   FromExcelToMap.java
21     public static void main(String[] args) {
22         Map<String,Integer> map = new TreeMap();
23         load(map, "data/Countries.dat");
24         print(map);
25         storeXL(map, "data/Countries.xls", "Countries Worksheet");
26     }
```

Listing 2-4 The FromMapToExcel program

Chapter 2

The load() method at line 23 loads the data from the Countries data file shown in *Figure 2-1* into the map. The print() method at line 24 prints the contents of the map. The storeXL() method at line 25 creates the Excel workbook as Countries.xls in our data folder, creates a worksheet named countries in that workbook, and then stores the map data into that worksheet.

The resulting Excel workbook and worksheet are shown in *Figure 2-6*.

Notice that the data is the same as in the file shown in *Figure 2-1*. The only difference is that, since the population of Brazil is over 100,000,000, Excel displays it as rounded and in exponential notation: 2.01E+08.

The code for the load() method is the same as that in Lines 15-26 of *Listing 2-1*, without line 16.

Here is the code for the print() method:

```
44      public static void print(Map map) {
45          Set countries = map.keySet();
46          for (Object country : countries) {
47              Object population = map.get(country);
48              System.out.printf("%-10s%,12d%n", country, population);
49          }
50      }
```

Listing 2-5 The print() method of the FromMapToExcel program

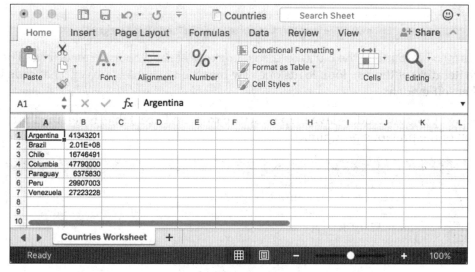

Figure 2-6 Excel workbook created by FromMapToExcel program

Data Preprocessing

In *Listing 2-5*, line 45 extracts the set of keys (countries) from the map. Then, for each of these, we get the corresponding population at line 47 and print them together at line 48.

Here is the code for the `storeXL()` method:

```java
class Country {
    protected String name;
    protected int population;
    protected int area;
    protected boolean landlocked;

    /* Constructs a new Country object from the next line being scanned.
       If there are no more lines, the new object's fields are left null.
    */
    public Country(Scanner in) {
        if (in.hasNextLine()) {
            this.name = in.next();
            this.population = in.nextInt();
            this.area = in.nextInt();
            this.landlocked = in.nextBoolean();
        }
    }

    @Override
    public String toString() {
        return String.format("%-10s %,12d %,12d %b",
                name, population, area, landlocked);
    }
}
```

Listing 2-6 The storeXL() method of the FromMapToExcel program

Lines 60-63 instantiate the `out`, `workbook`, `worksheet`, and `countries` objects. Then each iteration of the `for` loop loads one row into the `worksheet` object. That code is rather self-evident.

The next program loads a Java map structure from an Excel table, reversing the action of the previous program.

```java
public class FromExcelToMap {
    public static void main(String[] args) {
        Map map = loadXL("data/Countries.xls", "Countries Worksheet");
        print(map);
    }
}
```

Listing 2-7 The FromExcelToMap program

It simply calls this `loadXL()` method and then prints the resulting map:

```java
public static Map loadXL(String fileSpec, String sheetName) {
    Map<String,Integer> map = new TreeMap();
    try {
        FileInputStream stream = new FileInputStream(fileSpec);
        HSSFWorkbook workbook = new HSSFWorkbook(stream);
        HSSFSheet worksheet = workbook.getSheet(sheetName);
        DataFormatter formatter = new DataFormatter();
        for (Row row : worksheet) {
            HSSFRow hssfRow = (HSSFRow)row;
            HSSFCell cell = hssfRow.getCell(0);
            String country = cell.getStringCellValue();
            cell = hssfRow.getCell(1);
            String str = formatter.formatCellValue(cell);
            int population = (int)Integer.getInteger(str);
            map.put(country, population);
        }
    } catch (FileNotFoundException e) {
        System.err.println(e);
    } catch (IOException e) {
        System.err.println(e);
    }
    return map;
}
```

Listing 2-8 The loadXL() method of the FromExcelToMap program

The loop at lines 37-45 iterates once for each row of the Excel worksheet. Each iteration gets each of the two cells in that row and then puts the data pair into the map at line 44.

The code at lines 34-35 instantiates `HSSFWorkbook` and `HSSFSheet` objects. This and the code at lines 38-39 require the import of three classes from the external package `org.apache.poi.hssf.usermodel`; specifically, these three import statements:

```
import org.apache.poi.hssf.usermodel.HSSFRow;
import org.apache.poi.hssf.usermodel.HSSFSheet;
import org.apache.poi.hssf.usermodel.HSSFWorkbook;
```

The Java archive can be downloaded from https://poi.apache.org/download.html — POI-3.16. See *Appendix* for instructions on installing it in NetBeans.

XML and JSON data

Excel is a good visual environment for editing data. But, as the preceding examples suggest, it's not so good for processing structured data, especially when that data is transmitted automatically, as with web servers.

As an object-oriented language, Java works well with structured data, such as lists, tables, and graphs. But, as a programming language, Java is not meant to store data internally. That's what files, spreadsheets, and databases are for.

The notion of a standardized file format for machine-readable, structured data goes back to the 1960s. The idea was to nest the data in blocks, each of which would be marked up by identifying it with opening and closing tags. The tags would essentially define the grammar for that structure.

This was how the **Generalized Markup Language (GML)**, and then the **Standard Generalized Markup Language (SGML)**, were developed at IBM. SGML was widely used by the military, aerospace, industrial publishing, and technical reference industries.

The **Extensible Markup Language (XML)** derived from SGML in the 1990s, mainly to satisfy the new demands of data transmission on the World Wide Web. Here is an example of an XML file:

```xml
<?xml version="1.0" encoding="UTF-8"?>
<books>
    <book>
        <title>he Java Programming Language</title>
        <edition>4</edition>
        <author>Ken Arnold</author>
        <author>James Gosling</author>
        <author>David Holmes</author>
        <publisher>Addison Wesley</publisher>
        <year>2006</year>
        <isbn>0-321-34980-6</isbn>
    </book>
    <book>
        <title>Data Structures with Java</title>
        <author>John R. Hubbard</author>
        <author>JAnita Huray</author>
        <publisher>Prentice Hall</publisher>
        <year>2004</year>
        <isbn>0-313-093374-0</isbn>
    </book>
    <book>
        <title>Data Structures with Java</title>
        <author>John R. Hubbard</author>
        <publisher>Packt</publisher>
        <year>2017</year>
    </book>
</books>
```

Figure 2-7 An XML data file

This shows three `<book>` objects, each with a different number of fields. Note that each field begins with an opening tab and ends with a matching closing tag. For example, the field `<year>2017</year>` has opening tag `<year>` and closing tag `</year>`.

XML has been very popular as a data transmission protocol because it is simple, flexible, and easy to process.

The **JavaScript Object Notation (JSON)** format was developed in the early 2000s, shortly after the popularity of the scripting language JavaScript began to rise. It uses the best ideas of XML, with modifications to accommodate easy management with Java (and JavaScript) programs. Although the J in JSON stands for JavaScript, JSON can be used with any programming language.

There are two popular Java API libraries for JSON: `javax.jason` and `org.json`. Also, Google has a GSON version in `com.google.gson`. We will use the Official Java EE version, `javax.jason`.

JSON is a data-exchange format—a grammar for text files meant to convey data between automated information systems. Like XML, it is used the way commas are used in CSV files. But, unlike CSV files, JSON works very well with structured data.

In JSON files, all data is presented in name-value pairs, like this:

```
"firstName" : "John"
"age" : 54
"likesIceCream": true
```

These pairs are then nested, to form structured data, the same as in XML.

Figure 2-8 shows a JSON data file with the same structured data as the XML file in *Figure 2-7*.

```
Books.json
 1 {
 2      "books": [
 3          {
 4              "title": "The Java Programming Language",
 5              "edition": 4,
 6              "authors": [
 7                  "Ken Arnold",
 8                  "James Gosling",
 9                  "David Holmes"
10              ],
11              "publisher": "Addison Wesley",
12              "year": 2006,
13              "isbn": "0-321-34980-6"
14          },
15          {
16              "title": "Data Structures with Java",
17              "authors": [
18                  "John R. Hubbard",
19                  "Anita Huray"
20              ],
21              "publisher": "Prentice Hall",
22              "year": 2004,
23              "isbn": "0-13-093374-0"
24          },
25          {
26              "title": "Data Analysis with Java",
27              "author": "John R. Hubbard",
28              "publisher": "Packt",
29              "year": 2017
30          }
31      ]
32 }
```

Figure 2-8 A JSON data file

The root object is a name-value pair with `books` for its name. The value for that JSON object is a JSON array with three elements, each a JSON object representing a book. Notice that the structure of each of those objects is slightly different. For example, the first two have an `authors` field, which is another JSON array, while the third has a scalar `author` field instead. Also, the last two have no `edition` field, and the last one has no `isbn` field.

Each pair of braces { } defines a JSON object. The outer-most pair of braces define the JSON file itself. Each JSON object then is a pair of braces, between which are a string, a colon, and a JSON value, which may be a JSON data value, a JSON array, or another JSON object. A JSON array is a pair of brackets [], between which is a sequence of JSON objects or JSON arrays. Finally, a JSON data value is either a string, a number, a JSON object, a JSON array, `True`, `False`, or `null`. As usual, `null` means unknown.

JSON can be used in HTML pages by including this in the `<head>` section:

```
< script src =" js/ libs/ json2. js" > </ script >
```

If you know the structure of your JSON file in advance, then you can read it with a `JsonReader` object, as shown in *Listing 2-10*. Otherwise, use a `JsonParser` object, as shown in *Listing 2-11*.

A parser is an object that can read tokens in an input stream and identify their types. For example, in the JSON file shown in *Figure 2-7*, the first three tokens are {, books, and [. Their types are `START_OBJECT`, `KEY_NAME`, and `START_ARRAY`. This can be seen from the output in *Listing 2-9*. Note that the JSON parser calls a token an event.

```java
public class SortingData {
    public static void main(String[] args) {
        File file = new File("data/Countries.dat");
        TreeMap<Integer,String> dataset = new TreeMap();
        try {
            Scanner input = new Scanner(file);
            while (input.hasNext()) {
                String x = input.next();
                int y = input.nextInt();
                dataset.put(y, x);
            }
            input.close();
        } catch (FileNotFoundException e) {
            System.out.println(e);
        }
        print(dataset);
    }

    public static void print(TreeMap<Integer,String> map) {
        for (Integer key : map.keySet()) {
            System.out.printf("%,12d  %-16s%n", key, map.get(key));
        }
    }
}
```

```
run:
      6,375,830  Paraguay
     16,746,491  Chile
     27,223,228  Venezuela
     29,907,003  Peru
     41,343,201  Argentina
     47,790,000  Columbia
    201,103,330  Brazil
```

Listing 2-9 Identifying JSON event types

By identifying the tokens this way, we can decide what to do with them. If it's a `START_OBJECT`, then the next token must be a `KEY_NAME`. If it's a `KEY_NAME`, then the next token must be either a key value, a `START_OBJECT` or a `START_ARRAY`. And if that's a `START_ARRAY`, then the next token must be either another `START_ARRAY` or another `START_OBJECT`.

Data Preprocessing

This is called **parsing**. The objective is two-fold: extract both the key values (the actual data) and the overall data structure of the dataset.

```java
public class ParsingJSONFiles {
    public static void main(String[] args) {
        File dataFile = new File("data/Books.json");
        try {
            InputStream stream = new FileInputStream(dataFile);
            JsonParser parser = Json.createParser(stream);
            Event event = parser.next();   // advance past START_OBJECT
            HashMap<String,Object> map = getMap(parser);
            System.out.println(map);
            stream.close();
        } catch (FileNotFoundException e) {
            System.out.println(e);
        } catch (IOException e) {
            System.out.println(e);
        }
    }
}
```

Listing 2-10 Parsing JSON files

Here is the `getMap()` method:

```java
    /* Returns the HashMap parsed by the specified parser.
       Called when event.equals(event.START_OBJECT):
    */
    public static HashMap getMap(JsonParser parser) {
        HashMap<String,Object> map = new HashMap();
        Event event = parser.next();   // advance past START_OBJECT
        String key = parser.getString();
        event = parser.next();         // advance past KEY_NAME
        while (!event.equals(Event.END_OBJECT)) {
            if (event.equals(Event.VALUE_STRING)) {
                String value = parser.getString();
                map.put(key, value);
            } else if (event.equals(Event.VALUE_NUMBER)) {
                Integer value = parser.getInt();
                map.put(key, value);
            } else if (event.equals(Event.START_ARRAY)) {
                ArrayList<String> list = getList(parser);
                map.put(key, list);
            }
            event = parser.next();
            if (event.equals(Event.END_OBJECT)) {
                break;
            }
            key = parser.getString();
            event = parser.next();
        }
        return map;
    }
```

Listing 2-11 A getMap() method for parsing JSON files

And here is the `getList()` method:

```java
65        /* Returns the ArrayList parsed by the specified parser.
66           Called when event.equals(event.START_ARRAY):
67        */
68        public static ArrayList getList(JsonParser parser) {
          ArrayList list = new ArrayList();
70            Event event = parser.next();  // advance past START_ARRAY
71            while (!event.equals(Event.END_ARRAY)) {
                if (event.equals(Event.VALUE_STRING)) {
73                    list.add(parser.getString());
74                    event = parser.next();
75                } else if (event.equals(Event.START_OBJECT)) {
76                    HashMap<String,Object> map = getMap(parser);
77                    list.add(map);
78                    event = parser.next();
79                } else if (event.equals(Event.START_ARRAY)) {
80                    ArrayList subList = getList(parser);  // recursion
81                    list.add(subList);
82                    event = parser.next();
83                }
84            }
85            return list;
86        }
```

Listing 2-12 A getList() method for parsing JSON files

The actual data, both names and values, are obtained from the file by the methods `parser.getString()` and `parser.getInt()`.

Here is unformatted output from the program, just for testing purposes:

```
{books=[{year=2004, isbn=0-13-093374-0, publisher=Prentice Hall,
title=Data Structures with Java, authors=[John R. Hubbard, Anita
Huray]}, {year=2006, isbn=0-321-34980-6, edition=4, publisher=Addison
Wesley, title=The Java Programming Language, authors=[Ken Arnold,
James Gosling, David Holmes]}, {year=2017, author=John R. Hubbard,
publisher=Packt, title=Data Analysis with Java}]}
```

The default way that Java prints key-value pairs is, for example, `year=2004`, where `year` is the key and `2004` is the value.

Data Preprocessing

To run Java programs like this, you can download the file `javax.json-1.0.4.jar` from:

https://mvnrepository.com/artifact/org.glassfish/javax.json/1.0.4.

Click on **Download (BUNDLE)**.

See *Appendix* for instructions on how to install this archive in NetBeans.

Copy the downloaded jar file (currently `json-lib-2.4-jdk15.jar`) into a convenient folder (for example, `Library/Java/Extensions/` on a Mac). If you are using NetBeans, choose **Tools | Libraries** to load it into the IDE, and then right-click on the project icon and select **Properties** and then **Libraries**; choose **Add JAR/Folder** and then navigate to and select your **javax.json-1.0.4.jar** file.

Generating test datasets

Generating numerical test data is easy with Java. It boils down to using a `java.util.Random` object to generate random numbers.

```java
public class GeneratingTestData {
    private static final int ROWS = 8, COLS = 5;
    private static final Random RANDOM = new Random();

    public static void main(String[] args) {
        File outputFile = new File("data/Output.csv");
        try {
            PrintWriter writer = new PrintWriter(outputFile);
            for (int i = 0; i < ROWS; i++) {
                for (int j = 0; j < COLS-1; j++) {
                    writer.printf("%.6f,", RANDOM.nextDouble());
                }
                writer.printf("%.6f%n", RANDOM.nextDouble());
            }
            writer.close();
        } catch (FileNotFoundException e) {
            System.err.println(e);
        }
    }
}
```

Listing 2-13 Generating random numeric data

This program generates the following CSV file of eight rows and five columns of random decimal values.

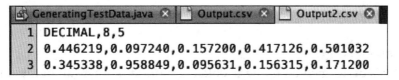

Figure 2-9 Test data file

Metadata

Metadata is data about data. For example, the preceding generated file could be described as *eight lines of comma-separated decimal numbers, five per line*. That's metadata. It's the kind of information you would need, for example, to write a program to read that file.

That example is quite simple: the data is unstructured and the values are all the same type. Metadata about structured data must also describe that structure.

The metadata of a dataset may be included in the same file as the data itself. The preceding example could be modified with a header line like this:

Figure 2-10 Test data file fragment with metadata in header

> When reading a data file in Java, you can scan past header lines by using the Scanner object's nextLine() method, as shown at line 32 in *Listing 2-15*.

Data cleaning

Data cleaning, also called **data cleansing** or **data scrubbing**, is the process of finding and then either correcting or deleting corrupted data values from a dataset. The source of the corrupt data is often careless data entry or transcription.

Various kinds of software tools are available to assist in the cleaning process. For example, Microsoft Excel includes a `CLEAN()` function for removing nonprintable characters from a text file. Most statistical systems, such as R and SAS, include a variety of more general cleaning functions.

Spellcheckers provide one kind of data cleaning that most writers have used, but they won't help against errors like `from` for `form` and `there` for `their`.

Statistical outliers are also rather easy to spot, for example, in our Excel `Countries` table if the population of Brazil appeared as `2.10` instead of `2.01E+08`.

Programmatic constraints can help prevent the entry of erroneous data. For example, certain variables can be required to have only values from a specified set, such as the ISO standard two-letter abbreviations for countries (CN for China, FR for France, and so on). Similarly, text data that is expected to fit pre-determined formats, such as phone numbers and email addresses, can be checked automatically during input.

An essential factor in the data cleaning process is to avoid conflicts of interest. If a researcher is collecting data to support a preconceived theory, any replacement of raw data should be done only in the most transparent and justifiable ways. For example, in testing a new drug, a pharmaceutical laboratory would maintain public logs of any data cleaning.

Data scaling

Data scaling is performed on numeric data to render it more meaningful. It is also called **data normalization**. It amounts to applying a mathematical function to all the values in one field of the dataset.

The data for Moore's Law provides a good example. *Figure 2-11* shows a few dozen data points plotted. They show the number of transistors used in microprocessors at various dates from 1971 to 2011. The transistor count ranges from 2,300 to 2,600,000,000. That data could not be shown if a linear scale were used for the transistor count field, because most of the points would pile on top of each other at the lower end of the scale. The fact is, of course, that the number of transistors has increased exponentially, not linearly. Therefore, only a logarithmic scale works for visualizing the data. In other words, for each data point (x, y), the point $(x, \log y)$ is plotted.

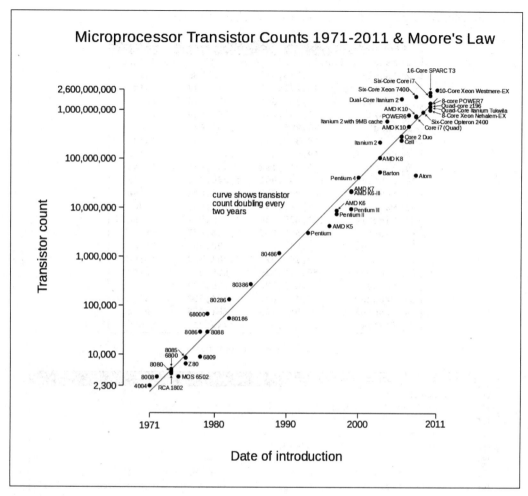

Figure 2-11 Moore's law

Microsoft Excel allows the use of scaling functions.

Data filtering

Filtering usually refers to the selection of a subset of a dataset. The selection would be made based upon some condition(s) on its data fields. For example, in a `Countries` dataset, we might want to select those landlocked countries whose land area exceeds 1,000,000 sq. km.

Data Preprocessing

For example, consider the `Countries` dataset shown in *Figure 2-12*:

```
Countries.dat
 1  Name         Population      Area       Landlocked
 2  Angola       26,655,513    1,245,585      false
 3  Argentina    44,272,125    2,732,847      false
 4  Bolivia      11,052,864    1,083,614      true
 5  Brazil      211,243,220    8,349,534      false
 6  Chad         14,965,482    1,257,604      true
 7  Chile        18,313,495      744,451      false
 8  Columbia     49,067,981    1,110,135      false
 9  Egypt        95,215,102      994,933      false
10  Ethiopia    104,344,901    1,000,430      false
11  Kenya        48,466,928      568,861      false
12  Mali         18,689,966    1,221,566      true
13  Nigeria     191,835,936      910,902      false
14  Niger        21,563,607    1,268,447      true
15  Paraguay      6,811,583      398,338      true
16  Peru         32,166,473    1,281,533      false
17  Tanzania     56,877,529      885,943      false
18  Uganda       41,652,938      199,774      true
19  Uruguay       3,456,877      174,590      false
20  Venezuela    31,925,705      881,926      false
21  Zambia       17,237,931      743,014      true
```

Figure 2-12 Data on Countries

```java
class Country {
    protected String name;
    protected int population;
    protected int area;
    protected boolean landlocked;

    /* Constructs a new Country object from the next line being scanned.
       If there are no more lines, the new object's fields are left null.
    */
    public Country(Scanner in) {
        if (in.hasNextLine()) {
            this.name = in.next();
            this.population = in.nextInt();
            this.area = in.nextInt();
            this.landlocked = in.nextBoolean();
        }
    }

    @Override
    public String toString() {
        return String.format("%-10s %,12d %,12d %b",
                name, population, area, landlocked);
    }
}
```

Listing 2-14 A class for data about Countries

For efficient processing, we first define the `Countries` class shown in *Listing 2-14*. The constructor at lines 20-27 read the four fields for the new `Country` object from the next line of the file being scanned by the specified `Scanner` object. The overridden `toString()` method at lines 29-33 returns a `String` object formatted like each line from the input file.

Listing 2-15 shows the main program to filter the data.

```java
public class FilteringData {
    private static final int MIN_AREA = 1000000;  // one million
    public static void main(String[] args) {
        File file = new File("data/Countries.dat");
        Set<Country> dataset = readDataset(file);

        for (Country country : dataset) {
            if (country.landlocked && country.area >= MIN_AREA) {
                System.out.println(country);
            }
        }
    }

    public static Set readDataset(File file) {
        Set<Country> set = new HashSet();
        try {
            Scanner input = new Scanner(file);
            input.nextLine();  // read past headers
            while (input.hasNextLine()) {
                set.add(new Country(input));
            }
            input.close();
        } catch (FileNotFoundException e) {
            System.out.println(e);
        }
        return set;
    }
}
```

Listing 2-15 Program to filter input data

Data Preprocessing

The `readDataset()` method at lines 28-41 uses the custom constructor at line 34 to read all the data from the specified file into a `HashSet` object, which is named `dataset` at line 19. The actual filtering is done at line 21. That loop prints only those countries that are landlocked and have an area of at least 1,000,000 sq. km., as shown in *Figure 2-13*.

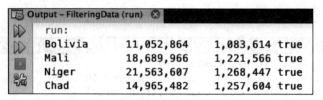

Figure 2-13 Filtered data

In Microsoft Excel, you can filter data by selecting **Data | Filter** or **Data | Advanced | Advanced Filter**.

Another type of data filtering is the process of detecting and removing noise from a dataset. In this context, noise refers to any sort of independent random transmission interference that corrupts the data. The term comes from the phenomenon of background noise in an audio recording. Similar phenomena occur with image files and video recordings. The methods for this kind of filtering are more advanced.

Sorting

Sometimes it is useful to sort or re-sort tabular data that is otherwise ready for processing. For example, the `Countries.dat` file in *Figure 2-1* is already sorted on its `name` field. Suppose that you want to sort the data on the `population` field instead.

One way to do that in Java is to use a `TreeMap` (instead of a `HashMap`), as shown in *Listing 2-16*. The `dataset` object, instantiated at line 17, specifies `Integer` for the key type and `String` for the value type in the map. That is because we want to sort on the `population` field, which has an integer type.

```java
public class SortingData {
    public static void main(String[] args) {
        File file = new File("data/Countries.dat");
        TreeMap<Integer,String> dataset = new TreeMap();
        try {
            Scanner input = new Scanner(file);
            while (input.hasNext()) {
                String x = input.next();
                int y = input.nextInt();
                dataset.put(y, x);
            }
            input.close();
        } catch (FileNotFoundException e) {
            System.out.println(e);
        }
        print(dataset);
    }

    public static void print(TreeMap<Integer,String> map) {
        for (Integer key : map.keySet()) {
            System.out.printf("%,12d   %-16s%n", key, map.get(key));
        }
    }
}
```

```
run:
   6,375,830   Paraguay
  16,746,491   Chile
  27,223,228   Venezuela
  29,907,003   Peru
  41,343,201   Argentina
  47,790,000   Columbia
 201,103,330   Brazil
```

Listing 2-16 Re-sorting data by different fields

The `TreeMap` data structure keeps the data sorted according to the ordering of its key field. Thus, when it is printed at line 29, the output is in increasing order of population.

Of course, in any map data structure, the key field values must be unique. So this wouldn't work very well if two countries had the same population.

Data Preprocessing

A more general approach would be to define a `DataPoint` class that implements the `java.util.Comparable` interface, comparing the objects by their values in the column to be sorted. Then the complete dataset could be loaded into an `ArrayList` and sorted simply by applying the `sort()` method in the `Collections` class, as `Collections.sort(list)`.

In Microsoft Excel, you can sort a column of data by selecting **Data | Sort** from the main menu.

Merging

Another preprocessing task is merging several sorted files into a single sorted file. *Listing 2-18* shows a Java program that implements this task. It is run on the two countries files shown in *Figure 2-12* and *Figure 2-13*. Notice that they are sorted by population:

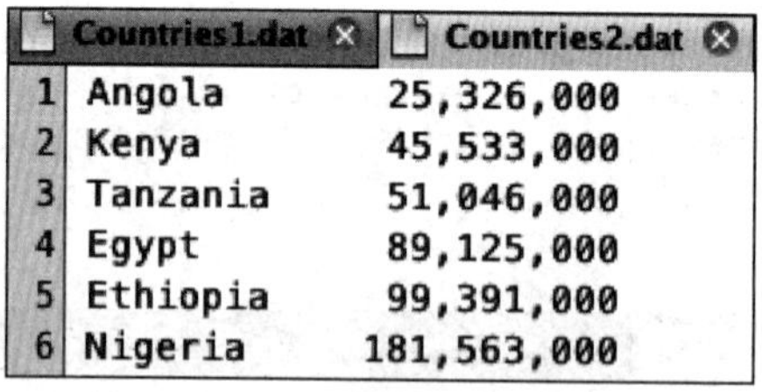

Countries1.dat	Countries2.dat
1 Angola	25,326,000
2 Kenya	45,533,000
3 Tanzania	51,046,000
4 Egypt	89,125,000
5 Ethiopia	99,391,000
6 Nigeria	181,563,000

Figure 2-12 African countries

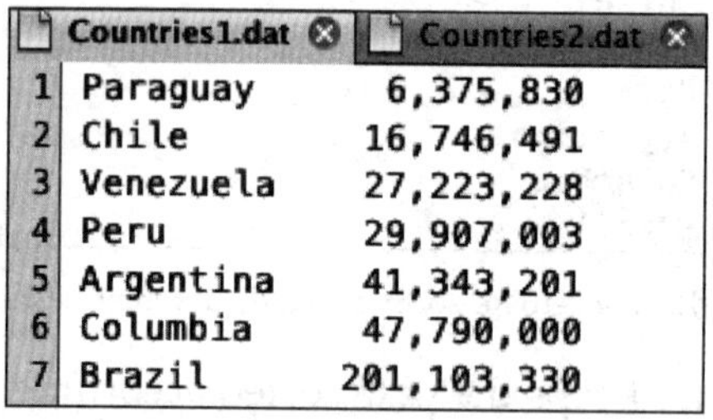

Countries1.dat	Countries2.dat
1 Paraguay	6,375,830
2 Chile	16,746,491
3 Venezuela	27,223,228
4 Peru	29,907,003
5 Argentina	41,343,201
6 Columbia	47,790,000
7 Brazil	201,103,330

Figure 2-13 South American countries

To merge these two files, we define a Java class to represent each data point, as shown in *Listing 2-17*. This is the same class as in *Listing 2-14*, but with two more methods added, at lines 30-38.

```
10  class Country implements Comparable {
11      private String name;
12      private Integer population;
13
14      public Country(String name, Integer population) {
15          this.name = name;
16          this.population = population;
17      }
18
19      /* Constructs a new Country object from the next line
20         of the specified file. If there are no more lines to read,
21         then the new object's fields are left null.
22      */
23      public Country(Scanner in) {
24          if (in.hasNextLine()) {
25              this.name = in.next();
26              this.population = in.nextInt();
27          }
28      }
29
30      public boolean isNull() {
31          return this.name == null;
32      }
33
34      @Override
35      public int compareTo(Object object) {
36          Country that = (Country)object;
37          return this.population - that.population;
38      }
39
40      @Override
41      public String toString() {
42          return String.format("%-10s%,12d", name, population);
43      }
44  }
```

Listing 2-17 Country class

By implementing the `java.util.Comparable` interface (at line 10), Country objects can be compared. The `compareTo()` method (at lines 34-38) will return a negative integer if the population of the implicit argument (this) is less than the population of the explicit argument. This allows us to order the Country objects according to their population size.

Data Preprocessing

The `isNull()` method at lines 30-32 is used only to determine when the end of the input file has been reached.

```java
13  public class MergingFiles {
14      public static void main(String[] args) {
15          File inFile1 = new File("data/Countries1.dat");
16          File inFile2 = new File("data/Countries2.dat");
17          File outFile = new File("data/Countries.dat");
18          try {
19              Scanner in1 = new Scanner(inFile1);
20              Scanner in2 = new Scanner(inFile2);
21              PrintWriter out = new PrintWriter(outFile);
22              Country country1 = new Country(in1);
23              Country country2 = new Country(in2);
24              while (!country1.isNull() && !country2.isNull()) {
25                  if (country1.compareTo(country2) < 0) {
26                      out.println(country1);
27                      country1 = new Country(in1);
28                  } else {
29                      out.println(country2);
30                      country2 = new Country(in2);
31                  }
32              }
33              while (!country1.isNull()) {
34                  out.println(country1);
35                  country1 = new Country(in1);
36              }
37              while (!country2.isNull()) {
38                  out.println(country2);
39                  country2 = new Country(in2);
40              }
41              in1.close();
42              in2.close();
43              out.close();
44          } catch (FileNotFoundException e) {
45              System.out.println(e);
46          }
47      }
48  }
```

Listing 2-18 Program to merge two sorted files

The program in *Listing 2-15* compares a Country object from each of the two files at line 24 and then prints the one with the smaller population to the output file at line 27 or line 30. When the scanning of one of the two files has finished, one of the Country objects will have null fields, thus stopping the while loop at line 24. Then one of the two remaining while loops will finish scanning the other file.

Chapter 2

```
MergingFiles.java   Countries.dat
1  Paraguay      6,375,830
2  Chile        16,746,491
3  Angola       25,326,000
4  Venezuela    27,223,228
5  Peru         29,907,003
6  Argentina    41,343,201
7  Kenya        45,533,000
8  Columbia     47,790,000
9  Tanzania     51,046,000
10 Egypt        89,125,000
11 Ethiopia     99,391,000
12 Nigeria     181,563,000
13 Brazil      201,103,330
```

Figure 2-14 Merged files

 This program could generate a very large number of unused Country objects.

For example, if one file contains a million records and the other file has a record whose population field is maximal, then a million useless (null) objects would be created. This reveals another good reason for using Java for file processing. In Java, the space used by objects that have no references is automatically returned to the heap of available memory. In a programming language that does not implement this garbage collection protocol, the program would likely crash for exceeding memory limitations.

Hashing

Hashing is the process of assigning identification numbers to data objects. The term **hash** is used to suggest a random scrambling of the numbers, like the common dish of leftover meat, potatoes, onions, and spices.

A good hash function has these two properties:

- **Uniqueness**: No two distinct objects have the same hash code
- **Randomness**: The hash codes seem to be uniformly distributed

Data Preprocessing

Java automatically assigns a hash code to each object that is instantiated. This is yet another good reason to use Java for data analysis. The hash code of an object, `obj`, is given by `obj.hashCode()`. For example, in the merging program in *Listing 2-15*, add this at line 24:

```
System.out.println(country1.hashCode());
```

You will get `685,325,104` for the hash code of the `Paraguay` object.

Java computes the hash codes for its objects from the hash codes of the object's contents. For example, the hash code for the string AB is `2081`, which is *31*65 + 66*— that is, 31 times the hash code for A plus the hash code for B. (Those are the Unicode values for the characters A and B.)

Of course, hash codes are used to implement hash tables. The original idea was to store a collection of objects in an array, `a[]`, where object x would be stored at index `i = h mod n`, where h is the hash code for x and n is the size of the array. For example, if `n = 255`, then the `Paraguay` object would be stored in `a[109]`, because 685,325,104 mod 255 = 109.

Recall that mod means remainder. For example, *25 mod 7 = 4* because *25 = 3·7 + 4*.

Summary

This chapter discussed various organizational processes used to prepare data for analysis. When used in computer programs, each data value is assigned a data type, which characterizes the data and defines the kind of operations that can be performed upon it.

When stored in a relational database, data is organized into tables, in which each row corresponds to one data point, and where all the data in each column corresponds to a single field of a specified type. The key field(s) has unique values, which allows indexed searching.

A similar viewpoint is the organization of data into key-value pairs. As in relational database tables, the key fields must be unique. A hash table implements the key-value paradigm with a hash function that determines where the key's associated data is stored.

Data files are formatted according to their file type's specifications. The comma-separated value type (CSV) is one of the most common. Common structured data file types include XML and JSON.

The information that describes the structure of the data is called its metadata. That information is required for the automatic processing of the data.

Specific data processes described here include data cleaning and filtering (removing erroneous data), data scaling (adjusting numeric values according to a specified scale), sorting, merging, and hashing.

3
Data Visualization

As the name suggests, this chapter describes the various methods commonly used to display data visually. A picture is worth a thousand words and a good graphical display is often the best way to convey the main ideas hidden in the numbers. Dr. Snow's Cholera Map (*Figure 1-3*) is a classic example.

Here is another famous example, also from the nineteenth century:

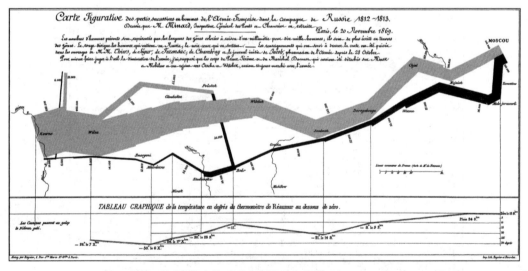

Figure 3-1. Minard's map of Napoleon's Russian campaign

This map shows the path of Napoleon and his army during the War of 1812, marching from France to Moscow and back. The main idea being conveyed here is that the size of the army at each point in the campaign is represented by the width of the lines of advance and retreat.

Tables and graphs

Most datasets are still maintained in tabular form, as in *Figure 2-12*, but tables with thousands of rows and many columns are far more common than that simple example. Even when many of the data fields are text or Boolean, a graphical summary can be much easier to comprehend.

There are several different ways to represent data graphically. In addition to more imaginative displays, such as Minard's map (*Figure 3-1*), we review the more standard methods here.

Scatter plots

A scatter plot, also called a scatter chart, is simply a plot of a dataset whose signature is two numeric values. If we label the two fields x and y, then the graph is simply a two-dimensional plot of those (x, y) points.

Scatter plots are easy to do in Excel. Just enter the numeric data in two columns and then select **Insert | All Charts | X Y (Scatter)**. Here is a simple example:

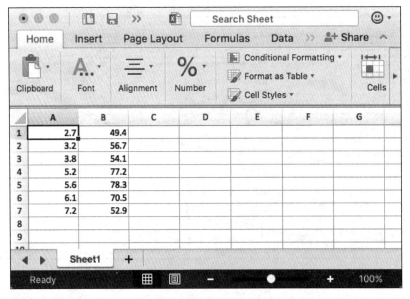

Figure 3-2. Excel data

The given data is shown in *Figure 3-2* and its corresponding scatter plot is in *Figure 3-3*:

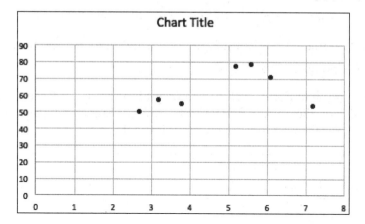

Figure 3-3. Scatter plot

The scales on either axis need not be linear. The microprocessor example in *Figure 2-11* is a scatter plot using a logarithmic scale on the vertical axis.

Figure 3-4 shows a scatter plot of data that relates the time interval between eruptions of the Old Faithful Geyser to the duration of eruption. This image was produced by a Java program that you can download from the Packt website for this book.

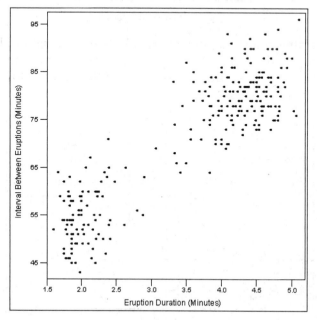

Figure 3-4. Scatter plot

Data Visualization

Line graphs

A line graph is like a scatter plot with these two differences:

- The values in the first column are unique and in increasing order
- Adjacent points are connected by line segments

To create a line graph in Excel, click on the **Insert** tab and select **Recommended Charts**; then click on the option: **Scatter with Straight Lines and Markers**. *Figure 3-5* shows the result for the previous seven-point test dataset.

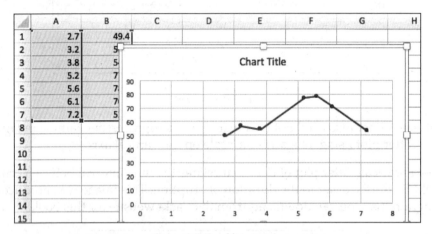

Figure 3-5. Excel line graph

The code for generating a line graph in Java is like that for a scatter plot. Use the `fillOval()` method of the `Graphics2D` class to draw the points and then use the `drawLine()` method. *Figure 3-6* shows the output of the `DrawLineGraph` program that is posted on the Packt website.

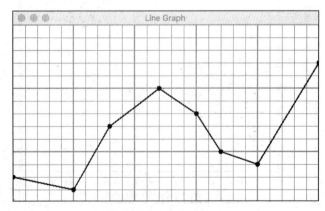

Figure 3-6. Java line graph

[48]

Bar charts

A bar chart is another common graphical method used to summarize small numeric datasets. It shows a separate bar (a colored rectangle) for each numeric data value.

To generate a bar chart in Java, use the `fillRect()` method of the `Graphics2D` class to draw the bars. *Figure 3-7* shows a bar chart produced by a Java program.

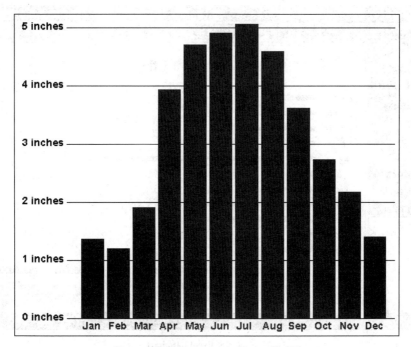

Figure 3-7. Precipitation in Knoxville TN

Bar charts are also easy to generate in Excel:

1. Select a column of labels (text names) and a corresponding column of numeric values.
2. On the **Insert** tab, click on **Recommended Charts**, and then either **Clustered Column** or **Clustered Bar**.

Data Visualization

The Excel bar chart in *Figure 3-8* shows a bar chart for the population column of our `AfricanCountries` dataset:

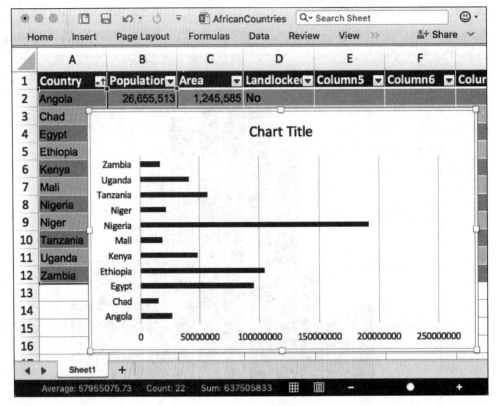

Figure 3-8. Excel bar chart

Histograms

A histogram is like a bar chart, except that the numeric data represent frequencies and so are usually actual numerical counts or percentages. Histograms are the preferred graphic for displaying polling results.

If your numbers add up to 100%, you can create a histogram in Excel by using a vertical bar chart, as in *Figure 3-9*.

Excel also has a plug-in, named **Data Analysis**, that can create histograms, but it requires numeric values for the category labels (called **bins**).

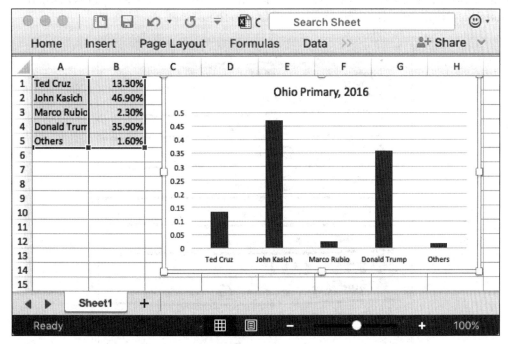

Figure 3-9. Excel histogram

Time series

A time series is a dataset in which the first field (the independent variable) is time. In terms of data structures, we can think of it as a map—a set of key-value pairs, where the key is time and the value represents an event that occurred at that time. Usually, the main idea is of a sequence of snapshots of some object changing with time.

Data Visualization

Some of the earliest datasets were time series. *Figure 3-10* shows a page of Galileo's 1610 notebook, where he recorded his observations of the moons of Jupiter. This is time series data: the time is written on the left, and the changing object is Galileo's sketch of the positions of the moons relative to the planet.

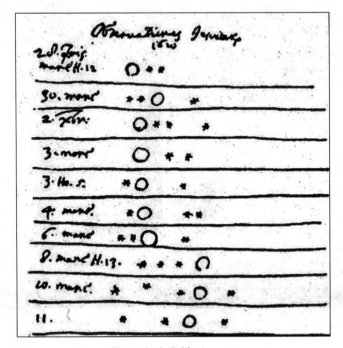

Figure 3-10. Galileo's notes

More modern examples of time series include biometric, weather, seismologic, and market data.

Most time series data are accumulated by automatic processes. Consequently, those datasets tend to be very large, qualifying as big data. This topic is explored in *Chapter 11, Big Data Analysis with Java*.

Digital audio and video files may also be regarded as time series datasets. Each sampling in the audio file or frame in the video file is the data value for an instant in time. But in these cases, the times are not part of the data because the intervals between them are all the same.

Java implementation

Listing 3-1 shows a test driver for a `TimeSeries` class. The class is parameterized, the parameter being the object type of the event that is moving through time. In this program, that type is `String`.

An array of six strings is defined at line 11. These will be the events to be loaded into the `series` object that is instantiated at line 14. The loading is done by the `load()`, defined at lines 27-31 and invoked at line 15.

```
TimeSeriesTester.java    TimeSeries.java
10 public class TimeSeriesTester {
11     final static String[] EVENTS = {"It", "was", "the", "best", "of", "times"};
12 
13     public static void main(String[] args) {
14         TimeSeries<String> series = new TimeSeries();
15         load(series);
16 
17         for (TimeSeries.Entry<String> entry : series) {
18             long time = entry.getTime();
19             String event = entry.getEvent();
20             System.out.printf("%16d: %s%n", time, event);
21         }
22 
23         ArrayList list = series.getList();
24         System.out.printf("list.get(3) = %s%n", list.get(3));
25     }
26 
27     static void load(TimeSeries<String> series) {
28         for (String event : EVENTS) {
29             series.add(System.currentTimeMillis(), event);
30         }
31     }
32 }
```

Listing 3-1. Test program for TimeSeries class

The contents of the series are loaded at line 15. Then at lines 17-21, the six key-value pairs are printed. Finally, at lines 23-24, we use the direct access capability of the `ArrayList` class to examine the series entry at index 3 (the fourth element). The output is shown in *Figure 3-11*.

```
Output - TimeSeriesTester (run)
   1492113428667: It
   1492113428669: was
   1492113428670: the
   1492113428672: best
   1492113428673: of
   1492113428674: times
list.get(3) = (1492113428672, best)
```

Figure 3-11. Output from the TimeSeriesTester program

Data Visualization

Notice that the list's element type is `TimeSeries.Entry`. That is a `static` nested class, defined inside the `TimeSeries` class, whose instances represent the key-value pairs.

The actual `TimeSeries` class is shown in *Listing 3-2*:

```java
public class TimeSeries<T> implements Iterable<TimeSeries.Entry> {
    private final Map<Long,T> map = new TreeMap();

    public void add(long time, T event) {
        map.put(time, event);
        try {
            TimeUnit.MICROSECONDS.sleep(1);   // 0.000001 sec delay
        } catch(InterruptedException e) {
            System.err.println(e);
        }
    }

    public T get(long time) {
        return map.get(time);
    }

    ArrayList getList() {
        ArrayList<TimeSeries.Entry> list = new ArrayList();
        for (TimeSeries.Entry entry : this) {
            list.add(entry);
        }
        return list;
    }

    public int size() {
        return map.size();
    }

    @Override
    public Iterator iterator() {...17 lines}

    public static class Entry<T> {...22 lines}
}
```

Listing 3-2. A TimeSeries class

The two folded code blocks at lines 43 and 61 are shown in *Listing 3-3* and *Listing 3-4*.

Line 15 shows that the time series data is stored as key-value pairs in a `TreeMap` object, with `Long` for the key type and `T` as the value type. The keys are `long` integers that represent time. In the test program (*Listing 3-1*), we used `String` for the value type `T`.

The `add()` method puts the specified time and event pair into the backing map at line 18, and then pauses for one microsecond at line 20 to avoid coincident clock reads. The call to `sleep()` throws an `InterruptedException`, so we have to enclose it in a `try-catch` block.

The `get()` method returns the event obtained from the corresponding `map.get()` call at line 27.

The `getList()` method returns an `ArrayList` of all the key-value pairs in the series. This allows external direct access by index number to each pair, as we did at line 23 in *Listing 3-1*. It uses a `for each` loop, backed by an `Iterator` object, to traverse the series (see *Listing 3-3*). The objects in the returned list have type `TimeSeries.Entry`, the nested class as shown in *Listing 3-4*.

```
43      public Iterator iterator() {
44          return new Iterator() { // anonymous inner class
45              private final Iterator it = map.keySet().iterator();
46
47              @Override
                public boolean hasNext() {
49                  return it.hasNext();
50              }
51
52              @Override
                public Entry<T> next() {
54                  long time = (Long)it.next();
55                  T event = map.get(time);
56                  return new Entry(time, event);
57              }
58          };
59      }
```

Listing 3-3. The iterator() method in the TimeSeries class

The `TimeSeries` class implements `Iterable<TimeSeries.Entry>` (*Listing 3-2*, line 14). That requires the `iterator()` method to be defined, as shown in *Listing 3-3*. It works by using the corresponding iterator defined on the backing map's key set (line 45).

Data Visualization

Listing 3-4 shows the `Entry` class, nested inside the `TimeSeries` class. Its instances represent the key-value pairs that are stored in the time series.

```java
    public static class Entry<T> {
        private final Long time;
        private final T event;

        public Entry(long time, T event) {
            this.time = time;
            this.event = event;
        }

        public long getTime() {
            return time;
        }

        public T getEvent() {
            return event;
        }

        @Override
        public String toString() {
            return String.format("(%d, %s)", time, event);
        }
    }
```

Listing 3-4. The nested entry class

 This `TimeSeries` class is only a demo. A production version would also implement `Serializable`, with `store()` and `read()` methods for binary serialization on disk.

Moving average

A moving average (also called **running average**) for a numeric time series is another time series whose values are averages of progressive sections of values from the original series. It is a *smoothing mechanism* to provide a more general view of the trends of the original series.

For example, the three-element moving average of the series (20, 25, 21, 26, 28, 27, 29, 31) is the series (22, 24, 25, 27, 28, 29). This is because 22 is the average of (20, 25, 21), 24 is the average of (25, 21, 26), 25 is the average of (21, 26, 28), and so on. Notice that the moving average series is smoother than the original, and yet still shows the same general trend. Also note that the moving average series has six elements, while the original series had eight. In general, the length of the moving average series will be $n - m + 1$, where n is the length of the original series and m is the length of the segments being averaged; in this case, $8 - 3 + 1 = 6$.

Listing 3-5 shows a test program for the `MovingAverage` class shown in *Listing 3-6*. The test is on the same preceding series: (20, 25, 21, 26, 28, 27, 29, 31).

```java
public class MovingAverageTester {
    static final double[] DATA = {20, 25, 21, 26, 28, 27, 29, 31};

    public static void main(String[] args) {
        TimeSeries<Double> series = new TimeSeries();
        for (double x : DATA) {
            series.add(System.currentTimeMillis(), x);
        }
        System.out.println(series.getList());

        TimeSeries<Double> ma3 = new MovingAverage(series, 3);
        System.out.println(ma3.getList());

        TimeSeries<Double> ma5 = new MovingAverage(series, 5);
        System.out.println(ma5.getList());
    }
}
```

Listing 3-5. Test program for the MovingAverage class

Some of the output for this program is shown in the following screenshot. The first line shows the first four elements of the given time series. The second line shows the first four elements of the moving average ma3 that averaged segments of length 3. The third line shows all four elements of the moving average ma5 that averaged segments of length 5.

```
Output - MovingAverageTester (run)
[(1492167782907, 20.0), (1492167782909, 25.0), (1492167782911, 21.0), (1492167782912, 26.0),
[(1492167782927, 22.0), (1492167782928, 24.0), (1492167782930, 25.0), (1492167782931, 27.0),
[(1492167782935, 24.0), (1492167782937, 25.4), (1492167782938, 26.2), (1492167782939, 28.2)]
```

Figure 3-12. Output from MovingAverageTester

Data Visualization

Here is the `MovingAverage` class:

```java
public class MovingAverage extends TimeSeries<Double> {
    private final TimeSeries parent;
    private final int length;

    public MovingAverage(TimeSeries parent, int length) {
        this.parent = parent;
        this.length = length;
        if (length > parent.size()) {
            throw new IllegalArgumentException("That's too long.");
        }

        double[] tmp = new double[length];   // temp array to compute averages
        double sum = 0;
        int i=0;
        Iterator it = parent.iterator();
        for (int j = 0; j < length; j++) {
            sum += tmp[i++] = nextValue(it);
        }
        this.add(System.currentTimeMillis(), sum/length);

        while (it.hasNext()) {
            sum -= tmp[i%length];
            sum += tmp[i++%length] = nextValue(it);
            this.add(System.currentTimeMillis(), sum/length);
        }
    }

    /* Returns the double value in the Entry currently located by it.
     */
    private static double nextValue(Iterator it) {
        TimeSeries.Entry<Double> entry = (TimeSeries.Entry)it.next();
        return entry.getEvent();
    }
}
```

Listing 3-6. A MovingAverage class

At line 1, the class `extends TimeSeries<Double>`. This means `MovingAverage` objects are specialized `TimeSeries` objects whose entries store numeric values of type `double`. (Remember that `Double` is just the wrapper class for `double`.) That restriction allows us to compute their averages.

The constructor (lines 14-35) does all the work. It first checks (at line 17) that the segment length is less than that of the entire given series. In our test program, we used segment lengths 3 and 5 on a given series of length 8. In practice, the length of the given series would likely be in the thousands. (Think of the daily Dow Jones Industrial Average.)

The `tmp[]` array defined at line 21 is used to hold one segment at a time as it progresses through the given series. The loop at lines 25-27 loads the first segment and computes the `sum` of those elements. Their average is then inserted as an entry into this `MovingAverage` series at line 28. That would be the pair (1492170948413, 24.0) in the preceding output from the `ma5` series.

The `while` loop at lines 30-34 computes and inserts the remaining averages. On each iteration, the `sum` is updated (at line 31) by subtracting the oldest `tmp[]` element and then (at line 32) replacing it with the next series value and adding it to the `sum`.

The `nextValue()` method at lines 39-42 is just a utility method to extract the numeric (`double`) value from the current element located by the specified iterator. It is used at lines 26 and 32.

It is easy to compute a moving average in Microsoft Excel. Start by entering the time series in two rows, replacing the time values with simple counting numbers (1, 2, …) like this:

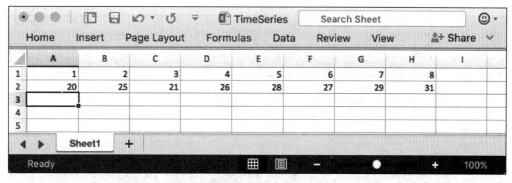

Figure 3-13. TimeSeries in Excel

This is the same data that we used previously, in *Listing 3-5*.

Now select **Tools | Data Analysis | Moving Average** to bring up the **Moving Average** dialog box:

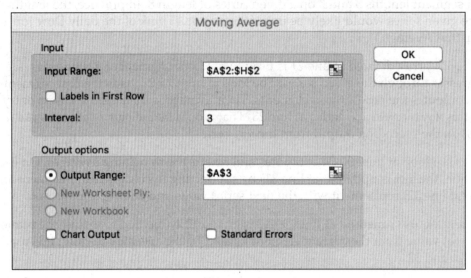

Figure 3-14. Excel's Moving Average dialog

Enter the **Input Range** (A2:A8), the **Interval** (3), and the **Output Range** (A3), as shown in the preceding screenshot. Then click **OK** to see the results:

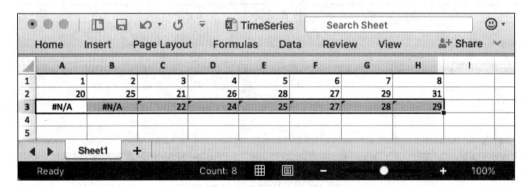

Figure 3-15. Excel Moving Average

These are the same results we got in *Figure 3-12*.

Figure 3-16 shows the results after computing moving averages with interval 3 and interval 5. The results are plotted as follows:

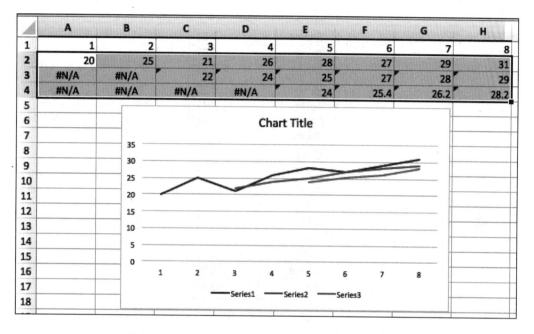

Figure 3-16. Plotted Moving Averages in Excel

Data ranking

Another common way to present small datasets is to rank the data, assigning ordinal labels (`first`, `second`, `third`, and so on) to the data points. This can be done by sorting the data on the key field.

Data Visualization

Figure 3-17 shows an Excel worksheet with data on students' grade-point averages (GPAs).

	A	B
1	Student	GPA
2	Adams	2.83
3	Baker	3.07
4	Cohen	3.61
5	Davis	2.49
6	Evans	3.11
7	Foley	2.72
8	Green	3.21
9	Haley	2.98
10	Irvin	3.14
11	Jones	2.05
12	Kelly	2.78
13	Lewis	3.29
14	Moore	3.67
15	North	2.75
16	Owens	2.93
17	Perry	3.61

Figure 3-17. Excel data

To rank these, select **Tools | Data Analysis | Rank and Percentile** to bring up the **Moving Average** dialog box:

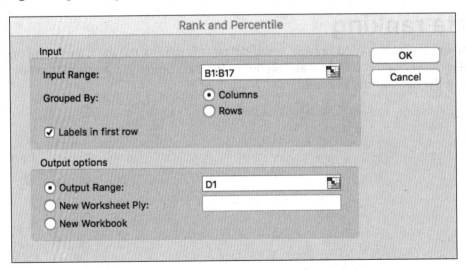

Figure 3-18. Ranking scores in Excel

Here, we have identified cells `B1` to `B17` as holding the data, with the first cell being a label. The output is to start in cell `D1`. The results are shown in the following screenshot:

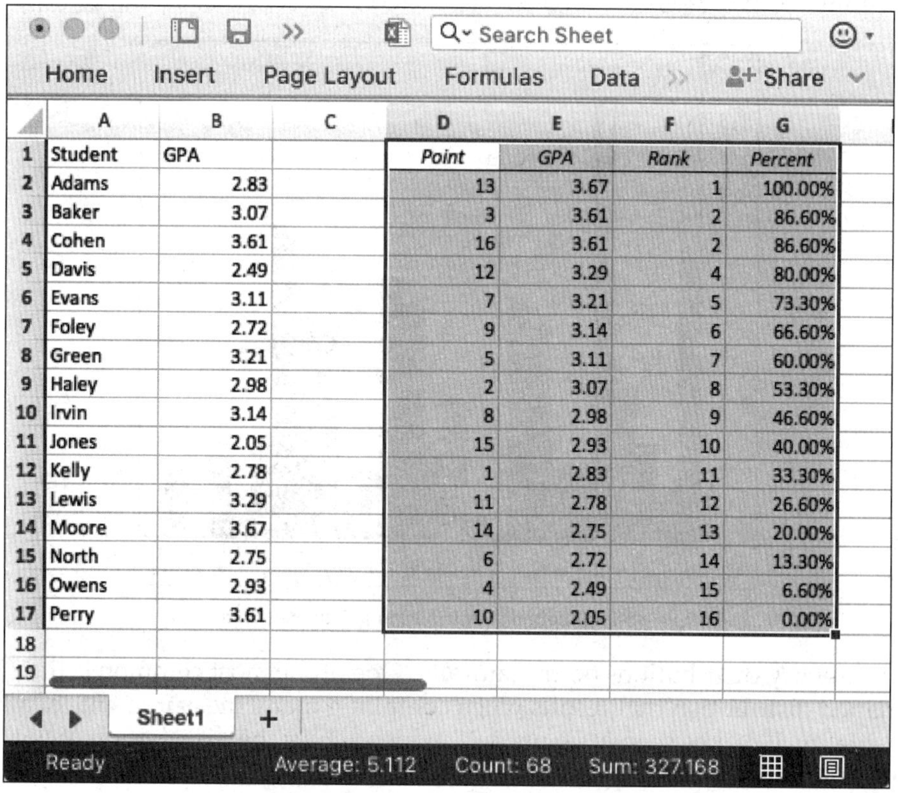

	A	B	C	D	E	F	G
1	Student	GPA		Point	GPA	Rank	Percent
2	Adams	2.83		13	3.67	1	100.00%
3	Baker	3.07		3	3.61	2	86.60%
4	Cohen	3.61		16	3.61	2	86.60%
5	Davis	2.49		12	3.29	4	80.00%
6	Evans	3.11		7	3.21	5	73.30%
7	Foley	2.72		9	3.14	6	66.60%
8	Green	3.21		5	3.11	7	60.00%
9	Haley	2.98		2	3.07	8	53.30%
10	Irvin	3.14		8	2.98	9	46.60%
11	Jones	2.05		15	2.93	10	40.00%
12	Kelly	2.78		1	2.83	11	33.30%
13	Lewis	3.29		11	2.78	12	26.60%
14	Moore	3.67		14	2.75	13	20.00%
15	North	2.75		6	2.72	14	13.30%
16	Owens	2.93		4	2.49	15	6.60%
17	Perry	3.61		10	2.05	16	0.00%
18							
19							

Figure 3-19. Results from Excel ranking

Column **D** contains the (relative) index of the record being ranked. For example, cells D3-G3 show the record of the third student in the original list (in column A) with name Cohen: that student ranks second, at the 86*th* percentile.

Frequency distributions

A **frequency distribution** is a function that gives the number of occurrences of each item of a dataset. It is like a histogram (for example, *Figure 3-8*). It amounts to counting the number or percentage of occurrences of each possible value.

Data Visualization

Figure 3-20 shows the relative frequency of each of the 26 letters in English language text. For example, the letter **e** is the most frequent, at 13%. This information would be useful if you were producing a word game like Scrabble or if you were attempting to decipher an encoded English message.

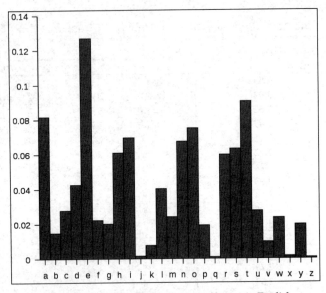

Figure 3-20. Frequency distribution of letters in English

Some frequency distributions occur naturally. One of the most common is the bell-shaped distribution that results when compiling many measurements of a single quality. It can be seen in *Figure 3-21*, which shows the distribution of the heights of American males, aged 25-34:

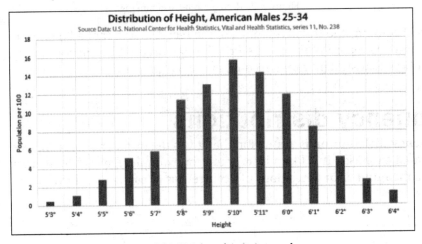

Figure 3-21. Heights of American males

In statistics, a dataset of measurements $\{x_1, x_2, ..., x_n\}$ is called a **sample**. The histogram shown in *Figure 3-20* comes from a large random sample of all American males. The x_i values would be in inches; for example, $x_{238} = 71$, for 5'11". From these numerical values are computed the sample mean, $\bar{X}$, and the sample standard deviation, s. The sample mean is the average x-value, and the sample standard deviation is a measure of how widely spread out the x-values are. These two parameters are computed by the formulas:

$$\bar{x} = \frac{1}{n}\sum_{i=1}^{n} x_i$$

$$s = \sqrt{\frac{1}{n-1}\sum_{i=1}^{n}(x_i - \bar{x})^2}$$

In this sample, $\bar{X}$ might be about 70.4 and s could be around 2.7.

The normal distribution

The normal distribution is a theoretical distribution that idealizes many distributions like the one in *Figure 3-20*. It is also called the bell shape curve or the Gaussian distribution, after its discoverer, Carl Friedrich Gauss (1777-1855).

The shape of the normal distribution is the graph of the following function:

$$f(x) = \frac{e^{-(x-\mu)^2/2\sigma^2}}{\sqrt{2\pi}\sigma}$$

Here, m is the mean and s is the standard deviation. The symbols e and π are a mathematical constant: $e = 2.7182818$ and $\pi = 3.14159265$. This function is called the **density function** for the (theoretical) distribution.

Note the distinction between the four symbols $\bar{X}$, s, m, and s. The first two are computed from the actual sample values; the second two are parameters used to define a theoretical distribution.

A thought experiment

To see how the normal distribution relates to actual statistics, imagine an experiment where you have a large flat clear jar that contains n (balanced) coins. When you shake the jar, some number x of those coins will settle heads up. The count x could be any number from 0 to n.

Now imagine repeating this experiment many times, each time recording the number x_i of heads that come up. For example, in the case of $n = 4$ coins, if you repeated the experiment 10 times, the sample could turn out to be {3, 2, 0, 3, 4, 2, 1, 2, 1, 3}. But, instead of only 10 times, suppose you did it 10,000 times. (Remember, this is just a thought experiment.) Then, after tabulating your results, you would form the corresponding histogram. That histogram would likely resemble the one at the top of *Figure 3-22*. (Remember, the coins are all supposed to be perfectly balanced.) There are five bars, one for each of the five x-values: 0, 1, 2, 3, and 4.

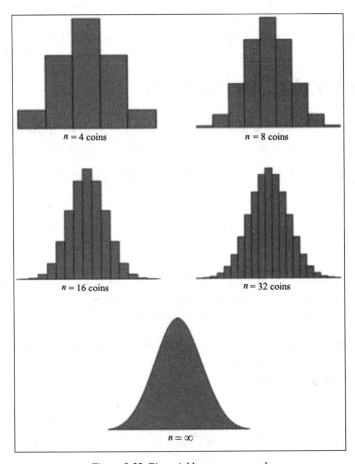

Figure 3-22. Binomial becomes normal

Next, add four more balanced coins to the jar, so now you have $n = 8$ coins. Repeat the process as before, and form that histogram. It would likely resemble the second histogram shown in *Figure 3-22*. The third and fourth histograms correspond to $n = 16$ coins and $n = 32$ coins. The fifth image is the histogram for $n = 1{,}024$ coins. It is also the graph of the normal distribution!

When phenomena can be attributed to a combination of many unbiased binary influences, such as 1,024 coins being flipped, their histograms tend to be normal. One can imagine the linear measure of any biological species, such as the height or the IQ of a population of males, as being so constituted.

The function $f(x)$ defined by the preceding formula is called the **probability density function** (**PDF**) for the normal distribution. Its graph describes the distribution: histograms for large samples of distributions that tend to be normal will look like that graph.

The PDF of a distribution can be used to compute probabilities. It is the areas under portions of the PDF that give these probabilities. For example, imagine in the preceding thought experiment the event of $23 \leq x \leq 26$, when $n = 32$. (Remember that x is the number of heads that come up.) Suppose you wanted to bet on that event happening. What is the probability? It is the ratio of the area of those four rectangles divided by the area of the entire histogram, but that area is the same area below the tops of those rectangles. If n is much larger, we could use the corresponding area under the graph of the PDF. This can all be done mathematically.

There is a famous anecdote told about the great English statistician R. A. Fischer and the formula for the normal distribution. He was once giving a public lecture about actuarial science, during which he wrote that formula on the chalkboard. A little old lady in the audience stopped him to ask the meaning of the Greek letter π. When Fisher replied, "That's the ratio of the circumference to the diameter of a circle", the lady exclaimed, "That can't be right—What possible connection could there be between circles and mortality?"

The exponential distribution

Of all the probability distributions, the normal (Gaussian) distribution is maybe be the most important, because it applies to so many common phenomena. The second most important is probably the exponential distribution. Its density function is as follows:

$$f(t) = \lambda e^{\lambda t}$$

Here, λ is a positive constant whose reciprocal is the mean ($\mu = 1$). This distribution models the time elapsed between randomly occurring events, such as radioactive particle emission or cars arriving at a toll booth. The corresponding **cumulative distribution function (CDF)** is as follows:

$$F(t) = 1 - e^{-\lambda t}$$

As an example, suppose that a university help desk gets 120 calls per eight-hour day, on average. That's 15 calls per hour, or one every four minutes. We can use the exponential distribution to model this phenomenon, with mean waiting time $\mu = 4$. That makes the density parameter $\lambda = 1/\mu = 0.25$, so:

$$F(t) = 1 - e^{-0.25t}$$

This means, for example, that the probability that a call comes in within the next five minutes would be:

$$F(5) = 1 - e^{-0.25(5)} = 1 - e^{-1.25} = 1 - 0.29 = 71\%$$

Java example

In *Figure 3-11*, we simulated a time series using random integers for the event times. To properly simulate events occurring at random times, we should instead use a process that generates timestamps whose elapsed time between events is exponentially distributed.

The CDF for any probability distribution is an equation that relates the probability $P = F(t)$ to the independent variable t. A simulation uses random numbers that represent probabilities. Therefore, to obtain the corresponding time t for a given random probability P, we must solve following the equation for t:

$$P = 1 - e^{-\lambda t}$$

That is:

$$t = -\frac{\ln(1-P)}{\lambda}$$

Here, $y = \ln(x)$ is the natural logarithm, which is the inverse of the exponential function $x = e^y$.

To apply this to our preceding Help Desk example, where $\lambda = 0.25$, we have:

$$t = -4\ln(1-P)$$

Note that, this time, t will be positive because the expression on the right is a double negative (*ln (1–P)* will be negative since $1 - P < 1$).

The program in *Listing 3-7* implements that formula at lines 14-17. At line 15, the `time()` method generates a random number p, which will be between 0 and 1. It then returns the result of the formula at line 16. The constant lambda is defined to be $\lambda = 0.25$ at line 12.

The output shows the eight numbers generated at line 21. Each of these represents a random inter-arrival time—that is, the time elapsed between randomly occurring events.

```java
public class ArrivalTimesTester {
    static final Random random = new Random();
    static final double LAMBDA = 0.25;

    static double time() {
        double p = random.nextDouble();
        return -Math.log(1 - p)/LAMBDA;
    }

    public static void main(String[] args) {
        for (int i = 0; i < 8; i++) {
            System.out.println(time());
        }
    }
}
```

```
run:
1.6242436905876936
1.5637305477120038
6.236197876991634
5.30829815924367
0.2435051295120049
15.591236886492794
0.8825444924230997
0.006795337507808347
```

Listing 3-7. *Simulating inter-arrival times*

The average of these eight numbers is 3.18, which is reasonably close to the expected value of 4.00 minutes between calls. Repeated runs of the program will produce similar results.

Summary

In this chapter, we have demonstrated various techniques for visualizing data, including the familiar scatter plots, line graphs, bar charts, and histograms. These were illustrated using the tools provided in Microsoft Excel. We also saw examples of time series data and computed their moving averages both in Excel and in Java.

We saw how visualizing (graphing) various binomial distributions can clarify how they lead to the normal distribution. We also saw, with a simple Java program, how the exponential distribution can be used to predict arrival times of random events.

4
Statistics

Statistics is what data science used to be called before the widespread use of computers. But, that use has not diminished the importance of statistical principles to the analysis of data. This chapter examines those principles.

Descriptive statistics

A **descriptive statistic** is a function that computes a numeric value which in some way summarizes the data in a numeric dataset.

We saw two statistics in *Chapter 3, Data Visualization*: the sample mean, $\bar{x}$, and the sample standard deviation, s. Their formulas are:

$$\bar{x} = \frac{1}{n}\sum_{i=1}^{n} x_i$$

$$s = \sqrt{\frac{1}{n-1}\sum_{i=1}^{n}(x_i - \bar{x})^2}$$

The mean summarizes the central tendency of the dataset. It is also called the **simple average** or **mean average**. The standard deviation is a measure of the dispersion of the dataset. Its square, s^2, is called the **sample variance**.

The **maximum** of a dataset is its greatest value, the **minimum** is its least value, and the **range** is their difference.

If **w** = $(w_1, w_2, ..., w_n)$ is a vector with the same number of components as the dataset, then we can use it to define the **weighted mean**:

$$\bar{x}_w = \frac{1}{n}\sum_{i=1}^{n} w_i x_i$$

In linear algebra, this expression is called the inner product of the two vectors, **w** and **x** = $(x_1, x_2, ..., x_n)$. Note that if we choose all the weights to be $1/n$, then the resulting weighted mean is just the sample mean.

The **median** of a dataset is the middle value, with the same number of values above as below. If the number of values is even, then the median is the mean of the two middle scores.

The **mode** is the most frequently occurring value in the dataset. This is a more specialized statistic, applying only to datasets that have many repeating values. But it also applies to non-numeric datasets. For example, in the frequency distribution of letters in *Figure 3-20*, the mode is the letter **e**.

The first, second, and third **quartiles** of a dataset are the numeric values at which 25%, 50%, and 75%, respectively, of the data values lie below. The **decile** and **percentile** statistics are defined similarly. Note that the mean is the same as the second quartile, the fifth decile, and the 50th percentile.

As an example, consider this set of student quiz scores:

$$S = \{9, 7, 9, 8, 5, 8, 6, 7, 8, 6\}$$

The mean is 7.3, the median is 7.5, the mode is 8, and the range is 4. The sample variance is 1.79 and the standard deviation is 1.34. If the teacher uses the weights (0.5, 0.5, 0.5, 1.0, 1.0, 1.0, 1.0, 1.5, 1.5, 1.5), counting the first three scores half as much as the middle scores and the last three 50% more, then the weighted mean for this dataset is $\bar{x}_w = 7.1$.

You can use Microsoft Excel to compute these descriptive statics, using its **Data Analysis** feature. Select **Tools | Data Analysis** and then **Descriptive Statistics** under **Analysis Tools**.

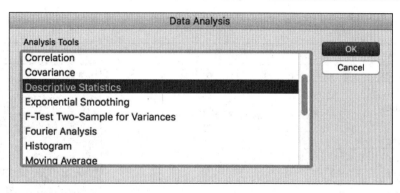

Figure 4-1. Excel's Data Analysis menu

To run this on the preceding quiz scores example, we entered the 10 scores in range B2:K2 in a worksheet, and then made the selections shown as follows:

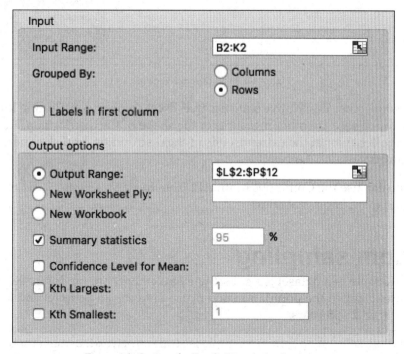

Figure 4-2. Options for Excel's Descriptive Statistics

The results can be seen in the following screenshot:

	A	B	C	D	E	F	G	H	I	J	K	L	M
2	Adams	9	7	9	8	5	8	6	7	8	6	Row1	
3	Baker												
4	Cohen											Mean	7.3
5	Davis											Standard Err	0.42295258
6	Evans											Median	7.5
7	Foley											Mode	8
8	Green											Standard De	1.33749351
9	Haley											Sample Varia	1.78888889
10	Irvin											Kurtosis	-0.8516537
11	Jones											Skewness	-0.3343605
12												Range	4
13												Minimum	5
14												Maximum	9
15												Sum	73
16												Count	10

Figure 4-3. Excel Descriptive Statistics

There are some other descriptive statistics that Excel computes, including **kurtosis** and **skewness**. These are other measures of the shape of the distribution. Kurtosis describes the behavior at the extreme ends of the distribution, and skewness refers to how asymmetric or unbalanced it is.

In Java, you can use the Commons Math API from http://apache.org/ to compute all these and many more statistics.

Random sampling

A **random sample** of a dataset is a subset whose elements are randomly selected. The given dataset is called the **population** and is usually very large; for example, all male Americans aged 25-35 years.

In a simulation, selecting a random sample is straightforward, using a random number generator. But in a real world context, random sampling is nontrivial.

Random sampling is an important part of quality control in nearly all types of manufacturing, and it is an essential aspect of polling in the social sciences. In industrial sectors such as pharmaceuticals, random sampling is critical to both production and testing.

To understand the principles of random sampling, we must take a brief detour into the field of mathematical probability theory. This requires a few technical definitions.

A **random experiment** is a process, real or imagined, that has a specified set of possible outcomes, any one of which could result from the experiments. The set S of all possible outcomes is called a **sample space**.

For example, the process of flipping four balanced coins. If the four coins are recognizably different, such as using a penny, a nickel, a dime, and a quarter, then we could specify the sample space as S_1 = {HHHH, HHHT, HHTH,..., TTTT}. The notation of an element such as HHTH means a head on the penny, a head on the nickel, a tail on the dime, and a head on the quarter. That set has 16 elements. Alternatively, we could use the same process without distinguishing the coins, say using four quarters, and then specify the sample space as S_2 = {0, 1, 2, 3, 4}, the number indicating how many of the coins turned up heads. These are two different possible sample spaces.

A **probability function** on a sample space S is a function p that assigns a number $p(s)$ to each element s of S, and is subject to these rules:

- $0 \leq p(s) \leq 1$, for every $s \in S$
- $\sum p(s) = 1$

For the first example, we could assign $p(s) = 1/16$ for each $s \in S_1$; that choice would make a good model of the experiment, if we assume that the coins are all balanced.

For the second example, a good model would be to assign the probabilities shown in the following table:

s	p(s)
0	1/16
1	1/4
2	3/8
3	1/4
4	1/16

Table 4-1. Probability distribution

As a third example, imagine the set S_3 of the 26 (lower case) letters of the Latin alphabet, used in English text. The frequency distribution shown in *Figure 3-20* provides a good probability function for that sample space. For example, $p("a") = 0.082$.

Statistics

The alphabet example reminds us of two general facts about probabilities:

- Probabilities are often expressed as percentages
- Although probabilities are theoretical constructs, they mirror the notion of relative frequencies

A probability function *p* assigns a number *p(s)* to each element *s* of a sample space S. From that, we can derive the corresponding **probability set function**: it assigns a number *P(U)* to each subset of the sample space S, simply by summing the probabilities of the subset's elements:

$$P(U) = \Sigma\{p(s) : s \in U\}$$

For example, in S_2, let *U* be the subset *U* = {3, 4}. This would represent the event of getting at least three heads from the four coins:

$$P(U) = \Sigma\{p(s) : s \in \{3,4\}\} = \Sigma\{p(3), p(4)\} = p(3) + p(4) = 1/4 + 1/16 = 5/16$$

Probability set functions obey these rules:

- $0 \leq P(U) \leq 1$, for every $U \subseteq S$
- $P(\emptyset) = 0$ and $P(S) = 1$
- $P(U \cup V) = P(U) + P(U) - P(U \cap V)$

The third rule can be seen from the following Venn diagram:

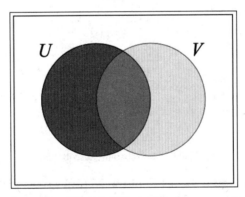

Figure 4-4. Venn diagram

The disk on the left represents the elements of U, and the right disk represents the elements of V. The values of P(U) and P(U) both include the values of P(U ∩ V), the green area, so those get counted twice. So, to sum the probabilities in all three regions only once each, we must subtract those that were counted twice.

A subset U of a sample space S is called an **event**, and the number P(U) represents the probability that the event will occur during the experiment. So, in this example, the probability of the event getting at least three heads is 5/16, or about 31%. In terms of relative frequencies, we could say that if we repeated the experiment many times, we should expect that event to be the outcome about 31% of the time.

Random variables

To understand statistical data analysis, we must first understand the concept of a random variable.

A **random variable** is a function that assigns a number to each element of a sample space. By converting symbolic outcomes such as HTTH, they allow for simpler mathematical analysis of the data.

The coin example illustrates this. The sample space S_1 has 16 outcomes. For each outcome x, let $X(x)$ be the number of heads in that outcome. For example, $X(HHTH) = 3$, and $X(TTTT) = 0$. This is the same as the transformation of S_1 into S_2 in the previous discussion. Here, we are using the random variable X for the same purpose. Now we can translate statements about the probability functions p and P into statements about purely numerical functions.

Probability distributions

The **probability distribution function (PDF)** that is induced by a random variable X is the function f_X, defined by:

$$f_X(x) = P(X = x), \text{ for each } x \text{ in the range } X(S)$$

Here, the expression $X = x$ means the event of all outcomes e for which $X(e) = x$.

Returning to our coin example, let's compute the probability distribution f_X. It is defined on the range of X, which is the set {0, 1, 2, 3, 4}. For example:

$$f_X(3) = P(X = 3) = P(\text{"3heads"})$$
$$= P(\{HHHT, HHTH, HTHH, THHH\}) = 4/16 = 1/4 = 0.25$$

Statistics

In fact, the probability distribution f_X for the first version of the coin example is precisely the same as the $p(s)$ function in the second version, tabulated in *Table 4-1*.

The properties of a probability distribution follow directly from those governing probabilities. They are:

- $0 \leq f(x) \leq 1$, for every $x \in X(S)$
- $\sum f(x) = 1$

Here is another classic example. The experiment is to toss two balanced dice, one red and one green, and observe the two numbers represented by the dots showing on top. The sample space S has 36 elements:

$$S = \{(1,1),(1,2),(1,3),\ldots,(6,5),(6,6)\}$$

If the dice are balanced, then each one of these 36 possible outcomes has the same probability: 1/36.

We define the random variable X to be the sum of the two numbers:

$$X(x_1, x_2) = x_1 + x_2$$

For example, $X(1, 3) = 4$, and $X(6, 6) = 12$.

The probability distribution f_X for X is shown in the following table:

x	$f_X(x)$
2	1/36
3	2/36
4	3/36
5	4/36
6	5/36
7	6/36
8	5/36
9	4/36
10	3/36
11	2/36
12	1/36

Table 4-2. Dice example

For example, $f_X(9) = 4/36$ because there are four elementary outcomes that satisfy the condition $x_1 + x_2 = 9$ and each of them has the probability 1/36.

We can see from its histogram in the following figure that this is a pretty simple distribution:

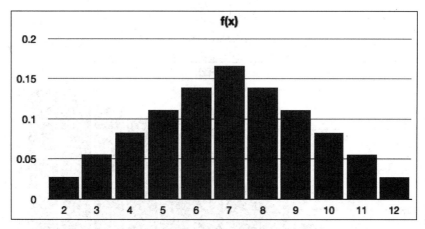

Figure 4-5. Dice probability distribution

Cumulative distributions

For every probability distribution function $f(x)$, there is a corresponding **cumulative distribution function** (**CDF**), denoted by $F(x)$ and defined as:

$$F(x) = \Sigma\{f(u) : u \leq x\}$$

The expression on the right means to sum all the values of $f(u)$ for $u \leq x$.

The CDF for the dice example is shown in *Table 4-3*, and its histogram is shown in *Figure 4-6*:

x	$f_X(x)$
2	1/36
3	3/36
4	6/36
5	10/36
6	15/36
7	21/36
8	26/36

[81]

Statistics

x	$f_X(x)$
9	30/36
10	33/36
11	35/36
12	36/36

Table 4-3. Dice example

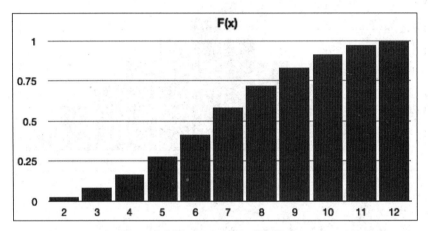

Figure 4-6. Dice cumulative distribution

The properties of a cumulative distribution follow directly from those governing probability distributions. They are:

- $0 \leq F(x) \leq 1$, for every $x \in X(S)$
- $F(x)$ is monotonically increasing; that is, $F(u) \leq F(v)$ for $u < v$
- $F(x_{max}) = 1$

Here, x_{max} is the maximum x value.

The CDF can be used to compute interval probabilities more easily that the PDF. For example, consider the event that $3 < X < 9$; that is, that the sum of the two dice is between 3 and 9. Using the PDF, the probability is computed as:

$$f(4) + f(5) + f(6) + f(7) + f(8) = 23/36$$

But computing it using the CDF is simpler:

$$F(8) - F(3) = 23/36$$

Of course, this assumes that the CDF (in *Table 4-3*) has already been generated from the PDF.

The binomial distribution

The binomial distribution is defined by this formula for its PDF:

$$f(x) = \binom{n}{x} p^x (1-p)^{n-x}, \text{ for } x = 0, 1, \ldots, n$$

Here, n and p are parameters: n must be a positive integer and $0 \leq p \leq 1$. The symbol $\binom{n}{x}$ is called a **binomial coefficient**. It can be computed from the following formula:

$$\binom{n}{x} = \frac{n!}{x!(n-x)!}$$

The exclamation point (!) stands for factorial, which means to multiply the integer by all its preceding positive integers. For example, five factorial is 5! = 5·4·3·2·1 = 120.

We encountered the binomial distribution in *Chapter 3, Data Visualization*, with the coin-flipping example. Here's a similar example. Suppose you have a bottle with five identical, balanced, tetrahedral dice. Each die has one face painted red and the other three faces painted green, as shown in the following figure:

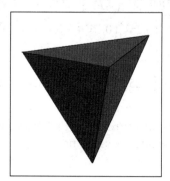

Figure 4-7. Tetrahedral Die

The experiment is to shake the flat-bottomed bottle and observe how the five dice land. Let X be the number of dice that land with a red face down. This random variable has a binomial distribution with $n = 5$ and $p = ¼ = 0.25$. Here is the PDF:

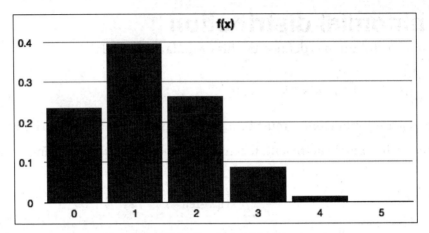

Figure 4-8. PDF for Tetrahedral Dice Experiment

Here are the calculations for the PDF:

$$f(0) = (1)(0.25^0)(0.75^5) = 243/1024 = 0.2373$$
$$f(1) = (5)(0.25^1)(0.75^4) = 405/1024 = 0.3955$$
$$f(2) = (10)(0.25^2)(0.75^3) = 270/1024 = 0.2637$$
$$f(3) = (10)(0.25^3)(0.75^2) = 90/1024 = 0.0879$$
$$f(4) = (5)(0.25^4)(0.75^1) = 15/1024 = 0.0146$$
$$f(5) = (1)(0.25^5)(0.75^0) = 1/1024 = 0.0010$$

The term **binomial coefficient** comes from the fact that these numbers result from the algebraic expansion of binomial expressions.

```
BinomialDistributionTester.java
 8  import org.apache.commons.math3.distribution.BinomialDistribution;
 9
10  public class BinomialDistributionTester {
        static final int n = 5;
        static final double p = 0.25;
13
14      public static void main(String[] args) {
15          BinomialDistribution bd = new BinomialDistribution(n, p);
16          for (int x = 0; x <= n; x++) {
17              System.out.printf("%4d%8.4f%n", x, bd.probability(x));
18          }
19          System.out.printf("mean = %6.4f%n", bd.getNumericalMean());
20          double variance = bd.getNumericalVariance();
21          double stdv = Math.sqrt(variance);
22          System.out.printf("standard deviation = %6.4f%n", stdv);
23      }
24  }
```

```
Output - BinomialDistributionTester (run)
    run:
       0   0.2373
       1   0.3955
       2   0.2637
       3   0.0879
       4   0.0146
       5   0.0010
    mean = 1.2500
    standard deviation = 0.9682
```

Listing 4-1. Binomial distribution demo

For example:

$$(p+q)^5 = 1p^5q^0 + 5P^4q^1 + 10p^3q^2 + 10p^2q^3 + 5p^1q^4 + 1p^0q^5$$

The coefficients are 1, 5, 10, 10, 5, and 1, which can be computed from $\binom{5}{x}$ using $x = 0, 1, 2, 3, 4,$ and 5.

Statistics

In general, the mean μ and standard deviation σ for any binomial distribution are given by the following formulas:

- $\mu = np$
- $\sigma = \sqrt{np(1-p)}$

```java
public class Simulation {
    static final Random RANDOM = new Random();
    static final int n = 5;         // number of dice used
    static final int N = 1000000;   // 1,000,000 simulations

    public static void main(String[] args) {
        double[] dist = new double[n+1];
        for (int i = 0; i < N; i++) {
            int x = numRedDown(n);
            ++dist[x];
        }
        for (int i = 0; i <= n; i++) {
            System.out.printf("%4d%8.4f%n", i, dist[i]/N);
        }
    }

    /* Simulates the toss of one tetrahedral die that has one red face and
        three green faces. Returns false unless the face down is red.
    */
    static boolean redDown() {
        int m = RANDOM.nextInt(4);   // 0 <= m < 4
        return (m == 0);             // P(m = 0) = 1/4
    }

    /* Simulates the toss of n tetrahedral dice that have one red face and
        three green faces. Returns the number that lands with red face down.
    */
    static int numRedDown(int n) {
        int numRed = 0;
        for (int i = 0; i < n; i++) {
            if (redDown()) {
                ++numRed;
            }
        }
        return numRed;
    }
}
```

```
run:
   0    0.2375
   1    0.3954
   2    0.2636
   3    0.0880
   4    0.0146
   5    0.0010
```

Listing 4-2. Simulation of binomial distribution

Thus, in our tetrahedral dice example, the mean is $(5)(0.25) = 1.25$, and the standard deviation is $\sqrt{(5)(0.25)(0.75)} = \sqrt{0.9375} = 0.9682$.

These calculations are verified by the output of the program shown in Listing 4-1. Note that the `BinaryDistribution` class provides the distribution's variance, but not its standard deviation. But the latter is just the square root of the former.

The program in Listing 4-2 simulates 1,000,000 runs of the experiment. The `redDown()` method at lines 26-32 simulates the toss of one tetrahedral die. The variable m is assigned 0, 1, 2, or 3 at line 30; all four values are equally likely, so there is a 25% chance it will be 0. Consequently, the methods returns `true` about 25% of the time.

The `numRedDown()` method at lines 34-45 simulates the tossing of the specified number (n) of tetrahedral dice. It counts the number of those dice that land red face down, and returns that count. That number is binomially distributed with $p = 0.25$. So when it is accumulated into the `dist[]` array at lines 16-20, the results should be very close to the theoretical values for $f(x)$ shown previously, as they are.

Multivariate distributions

A **multivariate probability distribution function** is one that is induced by several variables. It is also called a **joint probability function**.

For a simple example, take the two (six-sided) dice experiment again. But this time, suppose that one die is red and the other is green. Let X be the number that comes up on the red die and Y be the number on the green die. These are two random variables, each ranging from 1 to 6. Their probability distribution is a function of those two variables:

$$f_{XY}(x,y) = P(X = x \text{ and } Y = y)$$

For example:

$$f_{XY}(1,6) = P(X = 1 \text{ and } Y = 6) = P(\text{the red die is 1 and green die is 6}) = 1/36$$

You can see that this probability is 1/36 from the fact that there are 36 possible outcomes and they are all equally likely (assuming the dice are balanced).

Statistics

For a more interesting example, consider this experiment: a black bag contains two red marbles and four green marbles.

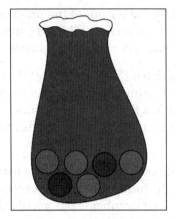

Figure 4-9. Bag of marbles

Except for their color, the marbles are identical. Two marbles are drawn at random from the bag, one after the other, without replacement. Let X be the number of red marbles from the first draw and Y the number from the second. Both variables are binary: the value is either 0 or 1. So:

$$f(0,0) = P(RR) = (2/6)(1/5) = 1/15 = 0.0667$$
$$f(0,1) = P(RG) = (2/6)(4/5) = 4/15 = 0.2667$$
$$f(1,0) = P(GR) = (4/6)(2/5) = 4/15 = 0.2667$$
$$f(1,1) = P(GG) = (4/6)(3/5) = 6/15 = 0.4000$$

The first calculation results from two out of six marbles being red on the first draw, and then one of the remaining five marbles being red on the second. The other three calculations are similar.

The joint probabilities are summarized in the following table:

$f(x,y)$	$Y = 0$	$Y = 1$
$X = 0$	1/15	4/15
$X = 1$	4/15	6/15

Table 4-4. Joint probabilities for marble experiment

Here is an example where relative frequencies are interpreted as probabilities. A Department of Motor Vehicles has data about the number of annual miles driven by drivers owning one or more vehicles. The random variables are X = number of vehicles owned, and Y = annual number of miles driven (per 5,000 miles). The following table shows all the relative frequencies:

	1 vehicle	2 vehicles	3 vehicles	
0-4,999 mi.	.09	.05	.02	.16
5,000-9,999 mi.	.18	.06	.05	.29
10,000-14,999 mi.	.13	.12	.11	.36
At least 15,000 mi.	.04	.08	.07	.19
	.44	.31	.25	1.00

Table 4-5. Percentages of drivers

For example, 9% of the drivers owned one vehicle and drove less than 5,000 miles.

The numbers in the margins summarize that row or column. For example, 16% of all drivers drove less than 5,000 miles, and 31% of all drivers owned two vehicles. These are called **marginal probabilities**.

Conditional probability

A **conditional probability** is one that is computed under the assumption that some related event is known to have occurred. For example, in the marble experiment, if we know that the first marble drawn was red, then the probability that the second marble is green is 4/5 = 80%. This is a conditional probability, the condition being that the first marble was red. It is written as:

$$P(2^{nd} \text{ is } G \mid 1^{st} \text{ is } R) = 4/5 = 80\%$$

(The vertical bar symbol | is read as "given").

On the other hand, the (unconditional) probability that the second marble is green is:

$$P(2^{nd} \text{ is } G) = P(RG) + P(GG) = 4/15 + 6/16 = 10/15 = 2/3 = 67\%$$

It should seem sensible that the probability that the second marble is green is greater (80%) after having removed a red marble. It should also seem sensible that the unconditional probability of green on the second (67%) is the same as the probability that the first is green (4/6).

Statistics

The general formula for the conditional probability of an event F, given an event E, is

$$P(F|E) = \frac{P(E \cap F)}{P(E)}$$

The symbolism $E \cap F$ means "E and F".

To apply this formula to our marbles example, let E be the event: *the first marble drawn is red* (1st is R) and let F be the event: *the second marble drawn is green* (2nd is G). Then:

$$P(F|E) = \frac{P(E \cap F)}{P(E)} = \frac{P(RG)}{P(PR) + P(RG)} = \frac{\left(\frac{4}{15}\right)}{\left(\frac{1}{15}\right) + \left(\frac{4}{15}\right)} = \frac{4}{5} = 80\%$$

If we cross-multiply by the denominator in the conditional probability formula, we get:

$$P(E \cap F) = P(E)P(F|E)$$

This has a simple interpretation: the probability that E and F both happen is the probability that E happens times the probability that F happens, given that E happened. That sounds like a temporal statement, suggesting that E happens before F. But mathematically, it works regardless of when the events might occur.

The independence of probabilistic events

We say that the two events E and F are **independent** if $P(F|E) = P(F)$. In other words, the occurrence of E has no effect upon the probability of F. From the previous formula, we can see that this definition is equivalent to the condition:

$$P(E \cap F) = P(E)P(F)$$

This shows that the definition is symmetric: E is independent of F if and only if F is independent of E.

In our preceding marble example, E = (1st is R) and let F = (2nd is G). Since $P(F|E)$ = 80% and $P(F)$ = 67%, we see that E and F are not independent. Obviously, F depends on E.

For another example, consider the previous Motor Vehicle example. Let E = (driver owns 2 vehicles) and F = (driver drives at least 10,000 miles/year). We can compute the unconditional probabilities from the marginal data: $P(E) = 0.31$ and $P(F) = 0.36 + 0.19 = 0.55$. So $P(E)P(F) = (0.31)(0.55) = 0.17$. But $P(E \cap F) = 0.12 + 0.08 = 0.20 \neq 0.17$, so these two events are not independent. Whether a person drives a lot depends upon the number of vehicles owned, according to this data.

Keep in mind here that when we speak of independence, we are referring to statistical independence, which is defined by the formula $P(F|E) = P(F)$; that is, it is defined entirely in terms of probabilities.

Contingency tables

Table 4-4 and *Table 4-5* are examples of **contingency tables** (also called **crosstab tables**). These are multivariate tabulations of two random variables in which one seeks to prove their dependence or independence.

Here is a classic example:

	Right-handed	Left-handed	
Male	43%	8%	51%
Female	45%	4%	49%
	88%	12%	100%

Table 4-6. Handedness by sex

In this population, 88% are right-handed. But 49% are female, and 45% are both female and right-handed, so 92% (0.45/0.49) of the females are right-handed. So being right-handed is not independent of sex.

Bayes' theorem

The conditional probability formula is:

$$P(F|E) = \frac{P(E \cap F)}{P(E)}$$

where E and F are any events (that is, sets of outcomes) with positive probabilities. If we swap the names of the two events, we get the equivalent formula:

$$P(E|F) = \frac{P(F \cap E)}{P(F)}$$

But $F \cap E = E \cap F$, so $P(F \cap E) = P(E \cap F) = P(F | E) P(E)$. Thus:

$$P(E|F) = \frac{P(F|E)P(E)}{P(F)}$$

This formula is called **Bayes' theorem**. The main idea is that it reverses the conditional relationship, allowing one to compute $P(E | F)$ from $P(F | E)$.

To illustrate Bayes' theorem, suppose the records of some Health Department show this data for 1,000 women over the age of 40 who have had a mammogram to test for breast cancer:

- 80 tested positive and had cancer
- 3 tested negative, but had cancer (a Type I error)
- 17 tested positive, but did not have cancer (a Type II error)
- 900 tested negative and did not have cancer

Notice the designations of errors of Type I and II. In general, a **Type I error** is when a hypothesis or diagnosis is rejected when it should have been accepted (also called a **false negative**), whereas a **Type II error** is when a conclusion is accepted when it should have been rejected (also called a **false positive**).

Here is a contingency table for the data:

	Positive	Negative	
Cancer	0.080	0.003	0.083
Benign	0.017	0.900	0.917
	0.097	0.903	1.000

Table 4-7. Cancer testing

If we think of conditional probability as a measure of a causal relationship, E causing F. Then we can use Bayes' theorem to measure the probability of the cause E.

For example, in the preceding contingency table, if we know what percentage of males are left-handed (P(L|M)), what percentage of people are male (P(M)), and what percentage of people are left-handed (P(L)), then we can compute the percentage of left-handed people that are male:

$$P(M|L) = \frac{P(L|M)P(M)}{P(L)}$$

Bayes' theorem is sometimes stated in this more general form:

$$P(F_k|E) = \frac{P(F_k)P(E|F_k)}{\sum_{i=1}^{n} P(F_i)P(E|F_i)}$$

Here, {$F_1, F_2, ..., F_n$} is a partition of the sample space into *n* disjoint subsets.

Covariance and correlation

Suppose X and Y are random variables with the joint density function *f(x, y)*. Then the following statistics are defined:

$$\mu_X = \sum_x \sum_y x f(x,y)$$

$$\mu_Y = \sum_x \sum_y y f(x,y)$$

$$\sigma_X^2 = \sum_x \sum_y (x - \mu_X)^2 f(x,y)$$

$$\sigma_Y^2 = \sum_x \sum_y (y - \mu_Y)^2 f(x,y)$$

$$\sigma_{XY} = \sum_x \sum_y (x - \mu_X)(y - \mu_Y) f(x,y)$$

$$\rho_{XY} = \frac{\sigma_{XY}}{\sigma_X \sigma_Y}$$

The statistic σ_{XY} is called the **covariance** of the bivariate distribution, and ρ_{XY} is called the **correlation coefficient**.

Statistics

It can be proved mathematically that if X and Y are independent, then $\sigma_{XY} = 0$. On the other hand, if X and Y are completely dependent (for example, if one is a function of the other), then $\sigma_{XY} = \sigma_X \sigma_Y$. These are the two extremes for covariance. Consequently, the correlation coefficient is a measure of the interdependence of X and Y, and is bounded as:

$$-1 \leq \rho_{XY} \leq 1$$

The program in *Listing 4-3* helps explain how the correlation coefficient works. It defines three two-dimensional arrays at lines 14-16. Each array has two rows, one for the X values and one for the Y values.

The first array, `data1`, is generated by the `random()` method, which is defined at lines 24-31. It contains 1,000 (x, y) pairs, the X-values in `data1[0]`, and the Y-values in `data1[1]`. The correlation coefficient for those pairs is printed at line 19. You can see from the output in *Figure 4-9* that $\rho_{XY} = 0.032$. That is nearly 0.0, indicating no correlation between X and Y, as one would expect with random numbers. We may conclude that X and Y are uncorrelated.

The `rho()` method, which returns the correlation coefficient ρ_{XY} is defined at lines 33-42. It implements the formulas shown previously. The variances σ_X^2 and σ_Y^2 are computed at lines 35 and 37, using the `Variance` object v, which is instantiated at line 34. Their corresponding standard deviations, σ_X and σ_Y are computed at lines 36 and 38. The covariance σ_{XY} is computed at line 40 from the covariance object c, which is instantiated at line 39. The `Covariance` and `Variance` classes are defined in the Apache Commons Math library, imported at lines 9-10.

```java
import java.util.Random;
import org.apache.commons.math3.stat.correlation.Covariance;
import org.apache.commons.math3.stat.descriptive.moment.Variance;

public class CorrelationExample {
    static final Random RANDOM = new Random();
    static double[][] data1 = random(1000);
    static double[][] data2 = {{1, 2, 3, 4, 5}, {1, 3, 5, 7, 9}};
    static double[][] data3 = {{1, 2, 3, 4, 5}, {9, 8, 7, 6, 5}};

    public static void main(String[] args) {
        System.out.printf("rho1 = %6.3f%n", rho(data1));
        System.out.printf("rho2 = %6.3f%n", rho(data2));
        System.out.printf("rho3 = %6.3f%n", rho(data3));
    }

    static double[][] random(int n) {
        double[][] a = new double[2][n];
        for (int i = 0; i < n; i++) {
            a[0][i] = RANDOM.nextDouble();
            a[1][i] = RANDOM.nextDouble();
        }
        return a;
    }

    static double rho(double[][] data) {
        Variance v = new Variance();
        double varX = v.evaluate(data[0]);
        double sigX = Math.sqrt(varX);
        double varY = v.evaluate(data[1]);
        double sigY = Math.sqrt(varY);
        Covariance c = new Covariance(data);
        double sigXY = c.covariance(data[0], data[1]);
        return sigXY/(sigX*sigY);
    }
}
```

Listing 4-3. Correlation coefficients

```
Output - CorrelationExample (run)
run:
rho1 =   0.032
rho2 =   1.000
rho3 =  -1.000
```

Figure 4-10. Correlation output

The second test set, data2, is defined at line 15. Its correlation coefficient, printed at line 20, is 1.000, indicating complete positive correlation. That is expected, since Y is a linear function of X: Y = 2X − 1. Similarly, the third test set, data3, defined at line 16, has complete negative correlation: Y = 10 − X, so ρ_{XY} = −1.000.

When ρ_{XY} is nearly 0.0, we may conclude that X and Y are uncorrelated. This means that we have no evidence of their linear correlation. It could be that the datasets are too small or too inaccurate to reveal such linear dependence. Or, they could be dependent in some other non-linear way.

The standard normal distribution

Recall from *Chapter 3, Data Visualization*, that the normal distribution's probability density function is:

$$f(x) = \frac{e^{-(x-\mu)^2/2\sigma^2}}{\sqrt{2\pi}\sigma}$$

where μ is the population mean and σ is the population standard deviation. Its graph is the well-known *bell curve*, centered at where $x = \mu$ and roughly covering the interval from $x = \mu-3\sigma$ to $x = \mu+3\sigma$ (that is, $x = \mu\pm3\sigma$). In theory, the curve is asymptotic to the x axis, never quite touching it, but getting closer as x approaches $\pm\infty$.

If a population is normally distributed, then we would expect over 99% of the data points to be within the $\mu\pm3\sigma$ interval. For example, the **American College Board Scholastic Aptitude Test in mathematics** (**AP math test**) was originally set up to have a mean score of $\mu = 500$ and a standard deviation of $\sigma = 100$. This would mean that nearly all the scores would fall between $\mu+3\sigma = 800$ and $\mu-3\sigma = 200$.

When $\mu = 0$ and $\sigma = 1$, we have a special case called the **standard normal distribution**.

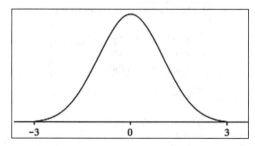

Figure 4-12. The standard normal distribution

Its PDF is:

$$\varphi(x) = \frac{e^{-x^2/2}}{\sqrt{2\pi}}$$

The normal distribution has the same general shape as the binomial distribution: symmetric, highest in the middle, shrinking down to zero at the ends. But there is one fundamental distinction between the two: the normal distribution is continuous, whereas the binomial distribution is discrete.

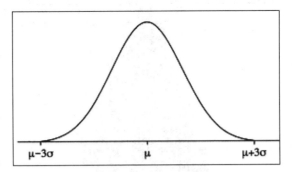

Figure 4-11. The normal distribution

A **discrete** distribution is one in which the range of the random variable is a finite set of values. A **continuous** distribution is one in which the range of the random variable is an entire interval of infinitely many values.

The images in *Figure 3-22* illustrate this distinction. The top histogram is the graph of the distribution of the number of heads in the flips of four fair coins; its range of values is the finite set {0, 1, 2, 3, 4}. The bottom image is the normal distribution; its range of values is the entire x axis—the interval $(-\infty, \infty)$.

When the distribution is discrete, we can compute the probability of an event simply by adding the probabilities of the elementary outcomes that form the event. For example, in the four-coin experiment, the probability of the event $E = (X > 2) = \{3, 4\}$ is $P(X > 2) = f(3) + f(4) = 0.25 + 0.06125 = 0.31125$. But when the distribution is continuous, events contain infinitely many values; they cannot be added in the usual way.

To solve this dilemma, look again at *Figure 3-22*. In each histogram, the height of each rectangle equals the elementary probability of the *x*-value that it marks. For example, in the top histogram (for the four-coin experiment), the heights of the last two rectangles are 0.25 and 0.06125, representing the probabilities of the outcomes $X = 3$ and $X = 4$. But if the width of each rectangle is exactly 1.0, then those elemental probabilities will also equal the areas of the rectangles: the area of the fourth rectangle is $A = (width)(height) = (1.0)(0.25) = 0.25$.

The histogram in the following figure shows the distribution of the number of heads from flipping 32 coins. That number X can be any integer from 0 to 32:

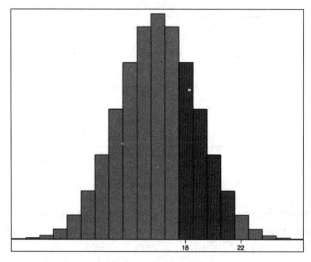

Figure 4-13. Flipping 32 coins

The four red rectangles represent the event E = {x : 17 < x ≤ 21} = {18, 19, 20, 21}. Note that the interval range uses "<" on the left and "≤" on the right. That ensures that the number of values in the range is equal to the difference of the two delimiting values:

$$21 - 17 = 4$$

The probability of this event, P(E), is the area of that red region:

$$P(E) = \Sigma\{f(x) : 17 < x \leq 21\}$$
$$= f(18) + f(19) + f(20) + f(21) = 0.1098 + 0.0809 + 0.0300 = 0.2733$$

So, the probability of an interval event like this is the area of the region that lies above the event interval and below the PDF curve. When the size of the range of X is small, we have to shift to the right a half unit, as in *Figure 4-13*, where we used the interval (17.5 ≤ x < 21.5) instead of (17 < x ≤ 21) for the base of the region. But as *Figure 3-22* suggests, a continuous random variable is nearly the same as a discrete random variable with a very large number of values in its range.

Instead of adding elementary probabilities to compute the probability of an event E, we can more simply take the difference of two values of the random variables' cumulative distribution function, its CDF. In the preceding example, that would be:

$$P(E) = \Sigma\{f(x) : 17 < x \leq 21\} = F(21) = 0.9749 - 0.7017 = 0.2733$$

This is how event probabilities are computed for continuous random variables.

The total area under the standard normal curve (*Figure 4-12*) is exactly 1.0. The cumulative distribution, CDF is denoted by Φ(x) and is computed from the formula:

$$\Phi(x) = \int_{-\infty}^{x} \varphi(u) \, du$$

This means the area under the standard normal curve φ and over the interval (−∞, x). That area equals the probability that X ≤ x. By subtracting areas, we have the formula for the probability that X is between the given numbers *a* and *b*:

$$P(a < X \leq b) = \Phi(b) - \Phi(a)$$

The normal curve in the following figure is the one that best approximates the binomial distribution that is shown in *Figure 4-13*:

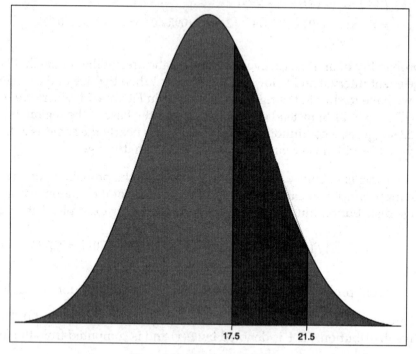

Figure 4-14. Normal probabilities

Its mean is $\mu = 16.0$ and its standard deviation is $\sigma = 2.82$. These values are computed from the formulas $\mu = np$ and $\sigma^2 = np(1-p)$. For this distribution, the area of the region is $P(17.5 < X \le 21.5)$, which is:

$$P(17.5 < X \le 21.5) = \Phi(21.5) - \Phi(17.5) = 0.941 - 0.7021 = 0.2720$$

We can see from the calculation for the binomial distribution that this approximation of 0.2720 for 0.2733 is quite good.

The program that computed these probabilities is shown in *Listing 4-4*:

BinomialDist.java | NormalDistributionTester.java

```
8  import org.apache.commons.math3.distribution.NormalDistribution;
9
10 public class NormalDistributionTester {
11     static int n = 32;
12     static double p = 0.5;
       static double mu = n*p;
       static double sigma = Math.sqrt(n*p*(1-p));
15
16     public static void main(String[] args) {
17         NormalDistribution nd = new NormalDistribution(mu, sigma);
18
19         double a = 17.5, b = 21.5;
20         double Fa = nd.cumulativeProbability(a);
21         System.out.printf("F(a) = %6.4f%n", Fa);
22         double Fb = nd.cumulativeProbability(b);
23         System.out.printf("F(b) = %6.4f%n", Fb);
24         System.out.printf("F(b) - F(a) = %6.4f%n", Fb - Fa);
25     }
26 }
```

Listing 4-4. Normal probabilities

It uses the `NormalDistribution` class from the Apache Commons Math library, as shown at line 8. Its `cumulativeProbability()` method returns the value of $\Phi(x)$.

Output – NormalDistributionTester (run)

```
run:
F(a) = 0.7021
F(b) = 0.9741
F(b) - F(a) = 0.2720
```

Figure 4-15. Output from Listing 4-4

The central limit theorem

A random sample is a set of numbers $S = \{x_1, x_2, \ldots, x_n\}$, each of which is a measurement of some unknown value that we seek. We can assume that each x_i is a value of a random variable X_i, and that all these random variables $X_1, X_2, \ldots, X_n$ are independent and have the same distribution with mean μ and standard deviation σ. Let S_n and Z be the random variables:

$$S_n = \sum_{i=1}^{n} X_i$$

$$Z = \frac{S_n - n\mu}{\sigma\sqrt{n}}$$

The central limit theorem states that the random variable Z tends to be normally distributed as n gets larger. That means that the PDF of Z will be close to the function $\varphi(x)$ and the larger n is, the closer it will be.

By dividing numerator and denominator by n, we have this alternative formula for Z:

$$Z = \frac{\frac{1}{n}S_n - \mu}{\sigma/\sqrt{n}}$$

This isn't any simpler. But if we designate the random variable $\bar{X}$ as:

$$\bar{X} = \frac{1}{n}\sum_{i=1}^{n} X_i = \frac{1}{n}S_n$$

then we can write Z as:

$$Z = \frac{\bar{X} - \mu}{\sigma/\sqrt{n}}$$

The central limit theorem tells us that this standardization of the random variable $\overline{X}$ is nearly distributed as the standard normal distribution $\varphi(x)$. So, if we take n measurements $x_1, x_2, \ldots, x_n$ of an unknown quantity that has an unknown distribution, and then compute their sample mean:

$$\overline{x} = \frac{1}{n}\sum_{i=1}^{n} x_i$$

we can expect this value

$$z = \frac{\overline{x} - \mu}{\sigma/\sqrt{n}}$$

to behave like the standard normal distribution.

Confidence intervals

The central limit theorem gives us a systematic way to estimate population means, which is essential to the quality control of automated production in many sectors of the economy, from farming to pharmaceuticals.

For example, suppose a manufacturer has an automated machine that produces ball bearings that are supposed to be 0.82 cm in diameter. The **quality control department (QCD)** takes a random sample of 200 ball bearings and finds that sample mean to be $\overline{x} = 0.824$ cm. From long-term previous experience, they have determined that machine's standard deviation s $\sigma = 0.042$ cm. Since $n = 200$ is large enough, we can assume that z is nearly distributed as the standard normal distribution, where:

$$z = \frac{\overline{x} - \mu}{\sigma/\sqrt{n}} = \frac{0.824 - \mu}{0.042/\sqrt{200}} = 336.7(0.824 - \mu)$$

Suppose that the QCD has a policy of 95% confidence, which can be interpreted as meaning that it tolerates error only 5% of the time. So their objective is to find an interval (a, b) within which we can be 95% confident that the unknown population mean μ lies; that is, $P(a \leq \mu \leq b) = 0.95$.

Using code like that in *Listing 4-4*, or from existing tabulations of the standard normal distribution, or from *Table 4-1*, we can find the number z_{95} for which $P(-z_{95} \leq z \leq z_{95}) = 0.95$. This constant z_{95} is called the **95% confidence coefficient**. It's value is $z_{95} = 1.96$; that is, $P(-1.96 \leq z \leq 1.96) = 0.95$.

Statistics

In this example, $z = 336.7(0.824 - \mu)$, so $\mu = 0.824 - z/336.7$, so $-1.96 \leq z \leq 1.96 \Leftrightarrow 0.818 \leq \mu \leq 0.830$. Thus, $P(0.818 \leq \mu \leq 0.830) = 0.95$. That is, we are 95% confident that the unknown population mean μ is between 0.818 cm and 0.830 cm.

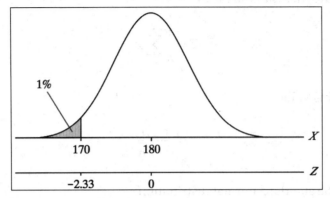

Figure 4-16. One-tailed critical region

Note that the interval bounds a and b were computed by these formulas:

$$a = \bar{x} - z_{95}\sigma / \sqrt{n}$$
$$b = \bar{x} - z_{95}\sigma / \sqrt{n}$$

This interval is called a **95% confidence interval**.

More generally, we can compute a confidence interval at any level c of confidence, simply by replacing z_{95} with z_c, where z_c satisfies the condition $P(-z_c \leq z \leq z_c) = c$. Several values of z_c are shown in *Table 4-8*. For example, $z_{99} = 2.58$. So, if the QCD in the preceding example wanted to elevate their quality policy to 99% confidence, they would use 2.58 in place of 1.96 to compute a 99% confidence interval:

$$a = \bar{x} - z_{99}\sigma / \sqrt{n} = 8.24 - (2.58)0.42/\sqrt{200} = 8.24 - 0.077 = 8.16$$
$$b = \bar{x} - z_{99}\sigma / \sqrt{n} = 8.24 - (2.58)0.42/\sqrt{200} = 8.24 + 0.077 = 8.32$$

So, the 99% confidence interval is (8.16, 8.32), meaning $\mu = 8.24 \pm 0.08$. As you would expect, that is less accurate that the 95% confidence interval we computed previously: (8.18, 8.30), meaning $\mu = 8.24 \pm 0.06$. The higher the confidence, the larger the confidence interval. The 100% confidence interval is $(-\infty, \infty)$.

Confidence Level, c%	Confidence Coefficient, z_c
99%	2.58
98%	2.33
96%	2.05
95%	1.96
90%	1.645
80%	1.28
68.25%	1.00

Table 4-8. Confidence Coefficients

Hypothesis testing

Suppose a pharmaceutical company claims that their allergy medicine is 90% effective in relieving allergies for a 12-hour period. To test that claim, an independent laboratory conducts an experiment with 200 subjects. Of them, only 160 report that the medicine was, as claimed, effective against allergies for 12 hours. The laboratory must determine whether that data is sufficient to reject the company's claim.

To set up the analysis, we first identify the population, the random sample, the relevant random variable, its distribution, and the hypothesis to be tested. In this case, the population could be all potential consumers of the medicine, the random sample is the set of $n = 200$ subjects reporting their results, and the random variable X is the number of those who did get the promised allergy relief. This random variable has the binomial distribution, with p being the probability that any one person does get that relief from taking the medicine. Finally, the hypothesis is that the medicine is (at least) 90% effective, which means that $p \geq 0.90$.

The hypothesis to be tested is traditionally called the **null hypothesis**, denoted as H_0. Its negation is called the **alternative hypothesis**, denoted as H_1.

We also specify the **level of significance** for the test. This is the probability threshold for erroneously rejecting the null hypothesis (called a **Type I error**). In this example, we choose the significance level to be $a = 1\%$. That is a rather standard choice among those who do these statistical tests. One way to interpret that choice is that the tester will be wrong in about 1% of tests where the null hypothesis was rejected.

So, here is our testing context:

- X is binomially distributed with $n = 200$ and p unknown
- $x = 160$
- $H_0: p = 0.90$
- $H_1: p < 0.90$
- $a = 0.01$

To complete the analysis, we assume that H_0 is true, and then use that value ($p = 0.90$) to try to determine whether the test results are unlikely; that is, that the probability of that event is less than the 1% significance level. That is called **rejecting the null hypothesis at the 1% level**. In that case, we would accept the alternative hypothesis H_1. If that probability is not less than the 1% threshold, we would then accept the null hypothesis H_0. Note that accepting H_0 does not prove it; it simply states that we do not have enough evidence to reject it. Nothing can be proved with statistics (unless, of course, the sample is the entire population).

The analysis depends upon the fact that, for large n, the binomial distribution is well approximated by the normal distribution. Here are the steps:

1. Determine the standard value z_1 for which $P(Z < z_1) = a$ for the standard normal distribution.
2. Compute $\mu = np$, $\sigma^2 = np(1 - p)$, and σ for this binomial distribution.
3. Compute x_1 from z_1, using $x_1 = \mu + z_1\sigma$.
4. Reject H_0 if $x < x_1$.

We can obtain z_1 from the following table. It shows that, for the standard normal distribution, $P(Z < -2.33) = 0.01$. So $z_1 = -2.33$.

Significance Level, a	Critical Value, z1
10%	±1.28
5%	±1.45
1%	±2.33
0.5%	±2.58
0.2%	±2.88

Table 4-9. Critical values for one-tailed testing

Now, we are assuming that $p = 0.9$ (the null hypothesis), so the mean is $\mu = np = (200)(0.9) = 180$, the variance is $\sigma^2 = np(1 - p) = (200)(0.9)(0.1) = 18$, and the standard deviation is $\sigma = 4.24$.

Next, $x_1 = \mu + z_1\sigma = 180 + (-2.33)(4.24) = 180 - 9.9 = 170$. This is the threshold x value. If H_0 is true, then it is very unlikely that X would be less than 170. Since the test value was $x = 160$, we must reject H_0. The company's "90% effective" claim is not supported by this data.

The hypothesis testing algorithm described previously applies only to **one-tailed tests**. That means that the alternative hypothesis is a single inequality, such as $p < 0.90$. In this case, the critical 1% region lies to the left of the critical value z_1, so we chose the negative value $z_1 = -2.33$. If instead, we had $H_1: p > 0.90$, then we would use the corresponding value $z_1 = 2.33$.

As an example of a **two-tailed test**, suppose the ball bearing manufacturer wanted to test the hypothesis that their ball bearings were 0.82 cm in diameter. This would be set up as:

$$H_0 : d = 0.82$$
$$H_1 : d \neq 0.82$$

In this case, there are two symmetric critical regions: one where $z < -z_1$ and the other where $z > z_1$.

Two-tailed hypothesis testing uses a different table than *Table 4-2*.

Summary

The field of statistics comprised the first organized study of data analysis. In this chapter, we have touched upon some of the main concepts in that field: random sampling, random variables, univariate and multivariate probability distributions, conditional probability and independence, Bayes' theorem and the central limit theorem, confidence intervals, and hypothesis testing. These are all topics that the data analyst should understand.

Some topics, such as the central limit theorem, are more mathematical and require some application of calculus for complete understanding. But even without that, we can appreciate the conclusion of the theorem, and we can recognize its operating principle when we see it.

5
Relational Databases

In *Chapter 2, Data Preprocessing*, we looked at some standard ways that data is stored. We saw that small unstructured datasets are often stored as text files, using white space, tabs, or commas to separate the data fields. Small, structured datasets are better handled by formats such as XML and JSON.

A database is a large, usually structured data collection that is accessed by an independent software system.

In this chapter, we will look at relational databases and the relational database systems that manage them. In *Chapter 10, Working with NoSQL Databases*, we will examine non-relational databases.

The relation data model

A **relational database** (**RDB**) is a database that stores its data in tables that are related by certain structural constraints. The word *relational* comes from the mathematical concept of a relation, which is essentially the same thing as a table. The precise definition follows.

A **domain** is a set of data values of the same data type, usually integer, decimal number, or text, but could be Boolean (`True`/`False`), nominal, or URL, among others. If $D_1, D_2, ..., D_n$ are domains, then their **Cartesian product** is the set $D_1, D_2, ..., D_n$ of all *n*-component sequences $t = (d_1, d_2, ..., d_n)$, where each $di \in Di$. Such sequences are called **tuples** (as in octuples when $n = 8$). A tuple is like a vector, except that the components of a tuple may be of different types; the components of a vector are usually just numbers. A **relation** is a subset of a Cartesian product of domains; that is, a set of tuples all from the same domain sequence $(D_1, D_2, ..., D_n)$, which is called the **schema** for the relation. We think of a relation as an ordinary table, where the rows are the tuples and the columns correspond to the domains.

Relational Databases

The following table shows an example of a relation whose domain schema could be integer, text, text, date, text, email. It has five tuples.

ID	Last Name	First Name	Date of Birth	Job Title	Email
49103	Adams	Jane	1975-09-02	CEO	jadams@xyz.com
15584	Baker	John	1991-03-17	Data Analyst	jbaker@xyz.com
34953	Cohen	Adam	1978-11-24	HR Director	acohen@xyz.com
23098	Davis	Rose	1983-05-12	IT Manager	rdavis@xyz.com
83822	Evans	Sara	1992-10-10	Data Analyst	sevans@xyz.com

Table 5-1. A database relation for employees

Each column has a unique name (for example, *Date of Birth*), which is shown at the top of the column when it is displayed. In addition, the table itself must have a name; this one is `Employees`.

Since tables and relations are practically the same thing, we will usually refer to them as tables, with the correspondingly simplified terms **rows** for its tuples, **columns** for its domains, attributes for its column names, and **fields** for its tuples' component values. A relational table is called an **instance** of its schema.

Note that a relation is a set of tuples, and as such, it is unordered. In other words, the order of the rows in a database table is irrelevant. The order of the columns is also irrelevant, since each column has a unique column name (for example, *Date of Birth*) and a type.

In addition to its domain schema, the definition of a table also usually includes a key designation. A **primary key** for a relation is a set of attributes whose values must be unique among the rows in the table. For *Table 5-1*, the obvious choice for the key attribute would be the *ID* field.

Relational databases

A relational database maintains a collection of database objects, including relations (tables) and the specified constraints upon them. The relation in the following table could be combined with the one in *Table 5-1* to form a relational database:

Dept ID	Name	Director
HR	Human Resources	34953
IT	Information Technology	23098
DA	Data Analysis	15584

Table 5-2. A database table of departments

This would also include the designated primary keys ID for the Employees table and Dept ID for the Departments table.

The **schema** for the relational database itself is the collection of table headings, their corresponding datatypes, and their key designations. We could specify the schema for this two-table database as:

- Employees(ID, Last Name, First Name, Date of Birth, Job Title, Email)
- Departments(Dept ID, Name, Director)

The primary keys are indicated by underlining their names. In this example, ID is the primary key for the Employees table, and Dept ID is the primary key for the Departments table. So, no two Employees records can have the same ID, and no two Departments records can have the same Dept ID.

Foreign keys

A **foreign key** in a database table is a field whose values are required to match corresponding primary key values, usually in another table. In the database defined previously, we would specify the foreign key Departments.Director to reference the primary key Employees.ID. So, for example, the *Director* of *Data Analysis* has the *ID 15584*, which identifies *John Baker* in *Table 5-1*.

Note that once a foreign key is designated in a table, no row may be added to it unless that key value matches an existing primary key value in the referenced table. Every row in the Departments table must have a Director value that matches an existing ID value in the Employees table.

When a foreign key in table *A* references a primary key in table *B*, we call *A* the child table and *B* the parent table. In this example, the Departments table is the child and the Employees table is the parent. Every child (Departments.Director) must have a parent (Employees.ID), and the parent must exist before the child exists. Note that this is a many-to-one relationship.

Foreign key constraints are one of the primary mechanisms that make relational databases efficient. They help minimize the duplication of data, which can be a major source of errors. For example, in the database described previously, without the power of the foreign key, we would have to replace the two tables with a single, substantially larger and less structured table.

Relational Databases

There are several important reasons why data duplication is bad in a database:

- If a data value is to be changed, every occurrence of it must be changed at the same time to maintain the consistency of the data. This is prone to error due to the possible difficulty of finding all those occurrences.
- A similar problem results when a data value is to be removed from the database.

The relational database model was developed by Edgar F. Codd while working at IBM in 1970. At the time, there were several competing database models.

Figure 5-1. Edgar F. Codd

But by the 1980s, the relational model had surpassed all others in popularity. Only with the advent of No-SQL databases in the last decade has the relational model lost any dominance in the field.

Relational database design

The design of a relational database is usually managed in several stages. The first stage is to identify the relational schema for the tables.

To clarify the process, consider the design of a small database of personally owned books. Here is a reasonable schema:

```
Authors(id, lastName, firstName, yob)
Books(title, edition, hardcover, publisher, pubYear, isbn, numPages)
Publishers(id, name, city, country, url)
```

Most of the fields specified here are text type (`String`, in Java). The `Authors.yob` (for year of birth), `Books.pubYear`, and `Books.numPages` fields should be integer type. The primary keys are `Authors.id`, `Books.isbn`, and `Publishers.id`. A publisher's ID would be a short string, such as A-W for Addison-Wesley, or Wiley for John Wiley and Sons.

The next step is to identify the foreign key constraints. One is obvious: `Books.publisher` references `Publishers.id`.

But we can't use foreign keys to link the `Authors` and `Books` tables because their relationship is many-to-many: an author could write several books, and a book may have several authors. Foreign key relationships must be many-to-one. The solution is to add a link table that uses foreign keys to connect to these other two tables: `AuthorsBooks(author, book)`.

This table has a two-field key: {`author`, `book`}. This is necessary because neither field by itself will have unique values. For example, it might contain these three rows:

Author	Book
JRHubb	978-0-07-147698-0
JRHubb	978-0-13-093374-0
AHuray	978-0-13-093374-0

Table 5-3. Link table for Library database

Now we can define the remaining foreign keys for our schema:

- `AuthorsBooks.author` references `Authors.id`
- `AuthorsBooks.book` references `Books.isbn`

Our complete schema now has four tables, each with a primary key, and three foreign keys:

- `Authors(id, lastName, firstName, yob)`
- `Books(title, edition, hardcover, publisher, pubYear, isbn, numPages)`
- `Publishers(id, name, city, country, url)`
- `AuthorsBooks(author, book)`
- `Books.publisher` references `Publishers.id`
- `AuthorsBooks.author` references `Authors.id`
- `AuthorsBooks.book` references `Books.isbn`

Relational Databases

This schema is illustrated in *Figure 5-2*. The primary keys are underlined, and the foreign keys are indicated by arrows.

Creating a database

Now we have decided upon our schema (*Figure 5-2*). It shows the four tables Authors, Books, Publishers, and AuthorsBooks. Each table lists its fields, with its primary key field(s) underlined. The foreign key constraints are shown by arrows. For example, Books.isbn is a foreign key for the AuthorsBooks.book field.

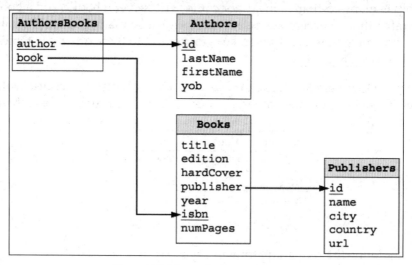

Figure 5-2. Schema for library database

The next step is to create the database. To do that, we must first choose a **relational database system (RDBMS)**.

There are many good RDBMSs available. If you have access to a proprietary RDBMS provided by your school or employer, that might be your best choice. Among the most popular are Oracle RDB, Microsoft SQL Server, and IBM DB2. There are also several good, open-access (free) RDBMSs available. The most popular among them is MySQL, which is described in the *Appendix*.

Since we have been using the NetBeans IDE for our Java examples in this book, we will use its Java DB database system in this chapter. However, all the SQL commands shown in the listings here should run the same in MySQL, Oracle, SQL Server, or DB2. The NetBeans Java DB website (https://netbeans.org/kb/docs/ide/java-db.html) provides more information on the Java DB database system. The website NetBeans MySQL explains how to connect to the MySQL database system from NetBeans.

Database developers communicate with a RDBMS by means of **SQL** (**Structured Query Language**). Unlike programming languages such as Java and Python, SQL is primarily a declarative language. That means that it responds to individual commands, like the language R or OS commands. Later in the chapter, we will see how to use JDBC to execute SQL commands from within Java programs.

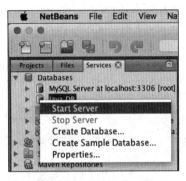

Figure 5-3. Starting Java DB

The RDBMS that comes with NetBeans is called **Java DB**. Follow these steps to create your `Library` database:

1. Click on the **Services** tab in the upper left panel.
2. Expand the **Databases** node.
3. Right-click on the **Java DB** node and select **Start Server** from the drop-down menu. The server should respond with a couple of messages in the output panel.
4. Right-click again on the **Java DB** node and this time select **Create Database…**.

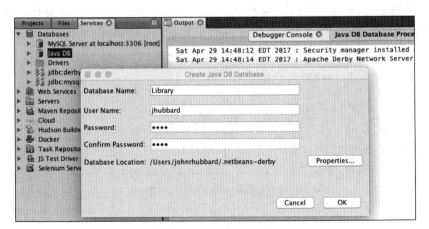

Figure 5-4. Creating the database

5. In the **Create Java DB Database** panel, enter `Library` for **Database Name**, and then pick a username and a simple password that you can remember. If you want to also specify the location for your database, click on the **Properties...** button.

6. A new node should now appear under **Databases**. Its label should be **jdbc:derby://localhost:1527/Library**. This represents your connection to the database. Its icon should look broken, indicating that the connection is currently closed.

7. Right-click on that connection icon and select **Connect....** This should cause the icon now to appear unbroken, thereby allowing you to expand that node.

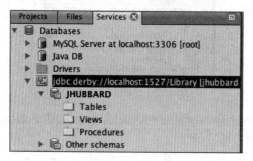

Figure 5-5. Connected to the database

8. Now right-click on your **Library** connection icon and select **Execute Command....** This will activate the SQL command editor, allowing you to enter and execute SQL commands.

9. Enter the `create table` command, exactly as shown in *Listing 5-1*. Notice the commas at the ends of the four lines 2-5, but not at the end of line 6 (in SQL, the comma is a separator, not a terminator). Also, in contrast to Java, SQL uses parentheses (lines 1 and 7) instead of braces to delimit the field definition list.

```
SQL 1 [jdbc:derby://localhost:15...]
1  create table Publishers (
2       id          char(4)      primary key,
3       name        varchar(32),
4       city        varchar(32),
5       country     char(2),
6       url         varchar(32)
7  );
```

Listing 5-1. SQL Create table command

10. Now right-click in the SQL editor window and select **Run Statement**. If there are no errors, the **Output** window will report that it executed successfully. Otherwise, go back and check your code against the one in *Listing 5-1*, correct it, and then run it again (you can also download this code from the Packt website). After it runs successfully, an icon for your new **PUBLISHERS** table will appear under the **Tables** node of your database, as shown in the following screenshot:

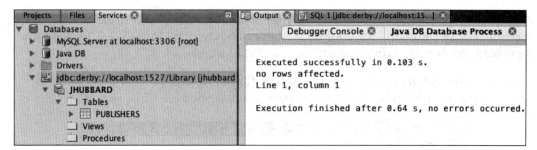

Figure 5-6. Publishers table created

SQL commands

The SQL query language is big and rather complex. This is due partly to its age—it's almost 50 years old. It's also partly because the language has evolved through many iterations, heavily influenced by powerfully competitive RDBMS vendors: IBM, Microsoft, and Oracle. But fear not; we'll be using only a small part of the language in this book.

SQL commands are separated into two groups: those that are used to define a database, and those that are used to access its data. The first is called the **Data Definition Language** (**DDL**), and the second is called the **Data Manipulation Language** (**DML**). The `create database` command used in *step 4* previously, and the `create table` command used in *Listing 5-1*, are DDL statements.

Here are some other common DDL commands:

    alter table
    drop table
    create index

Here are a few common DML commands:

    insert into
    update
    delete from

> We use the terms command and statement interchangeably in SQL.
>
> Many authors prefer to write SQL statements using all capital letters. But unlike Java and most other procedural languages, SQL is case insensitive—it ignores differences between uppercase and lowercase. Whether you use uppercase or lowercase is mainly just a question of style.

Continuing the creation of your `Library` database, the next step is to execute the `create table` command for your `Books` table, as shown in *Listing 5-2*. This has some new features that we didn't have in the `Publishers` table.

First, note that we are using the `int` datatype (lines 3, 6, and 8). This is the same as the `int` type in Java. In SQL, it can also be written as `integer`.

```
create table Books (
    title      varchar(64),
    edition    int          default 1,
    cover      char(4)      check(cover in ('HARD','SOFT')),
    publisher  char(4)      references Publishers,
    pubYear    int,
    isbn       char(13)     primary key,
    numPages   int
);
```

Listing 5-2. Create books command

Secondly, in the declaration of the `cover` field (line 4), we are using a `check` clause to limit the possible values of that field to only `'HARD'` and `'SOFT'`. This is essentially a nominal datatype, comparable to an enumerated (`enum`) type in Java.

Also note that in SQL, all strings are delimited by apostrophes (the single quote character) instead of regular quotation marks as in Java: write `'HARD'` not `"HARD"`.

Finally, note the `references` clause in line 5. This enforces the foreign key: `Books.publisher` references `Publishers.id`. Notice that both fields, `Books.publisher` and `Publishers.id`, have the same datatype: `char(4)`. A foreign key should have the same datatype as the primary key that it references.

Your database structure should now look like the display in the following screenshot, but with your username in place of **JHUBBARD**:

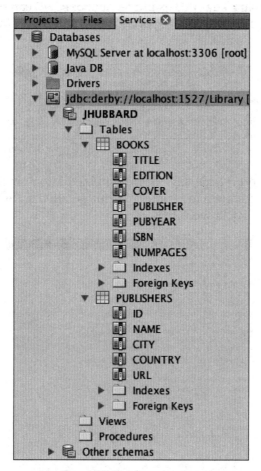

Figure 5-7. Database structure

The SQL commands for creating the `Authors` table and the `AuthorsBooks` table are shown in *Listing 5-3* and *Listing 5-4*:

```
create table Authors (
    id          char(8)      primary key,
    lastName    varchar(16),
    firstName   varchar(16),
    yob         int
)
```

Listing 5-3. Create Authors command

In *Listing 5-4*, the two required foreign keys are specified by `references` clauses (lines 2 and 3). Also note that the two-field `primary key` is specified separately, at line 4. Single-field primary keys (for example, `Authors.id`) can be specified either way, as an appending clause or on a separate line. But multi-field keys must be specified separately this way.

```
1  create table AuthorsBooks (
2      author    char(8)     references Authors,
3      book      char(13)    references Books,
4      primary key(author, book)
5  )
```

Listing 5-4. Create AuthorsBooks

All four of these SQL commands have been combined into a single file, named `CreateTables.sql` and shown in *Listing 5-5*. This is called an **SQL script**, which you can download from the Packt website.

```
1  ----------------------------------------------------------
2  --  CreateTables.sql
3  --  Creates four tables for the Library database.
4  --  Data Analysis with Java
5  --  John R. Hubbard
6  --  May 4 2017
7  ----------------------------------------------------------
8
9  drop table AuthorsBooks;
10 drop table Authors;
11 drop table Books;
12 drop table Publishers;
13
14 create table Publishers (
15     id         char(4)        primary key,
16     name       varchar(32),
17     city       varchar(32),
18     country    char(2),
19     url        varchar(32)
20 );
21
22 create table Books (
23     title      varchar(64),
24     edition    int            default 1,
25     cover      char(4)        check(cover in ('HARD','SOFT')),
26     publisher  char(4)        references Publishers,
27     pubYear    int,
28     isbn       char(13)       primary key,
29     numPages   int
30 );
31
32 create table Authors (
33     id         char(8)        primary key,
34     lastName   varchar(16)    not null,
35     firstName  varchar(16),
36     yob        int            default 0
37 );
38
39 create table AuthorsBooks (
40     author     char(8)        references Authors,
41     book       char(13)       references Books,
42     primary key (author, book)
43 );
```

Listing 5-5. SQL script creating all four tables

The script begins with a seven-line header comment that identifies the file by name and states its purpose, identifies the book, the author, and the date it was written. In SQL, anything that appears on a line after a double dash is ignored by the SQL interpreter as a comment.

The four tables must first be deleted before they can be recreated. This is done with the `drop table` command at lines 9-12.

The tables must be dropped in the reverse order from which they were created. This accommodates the foreign key constraints. For example, the `Books` table depends on the `Publishers` table (see *Figure 5-2*), so it must be dropped before the `Publishers` table. Of course, the order for the creations of the tables is the reverse, for the same reason: The `Publishers` table must be created before the `Books` table is.

Notice that this script contains eight SQL statements: four `drop` and four `create`. The statements must be separated by semicolons, although a single statement can be run without a semicolon.

To run the script in the NetBeans SQL editor, right-click anywhere in the window and select **Run File**.

Inserting data into the database

Data is inserted into a database table by means of the `insert into` statement. This is illustrated in the following screenshot:

```
1  insert into Publishers values (
2      'PPL',
3      'Packt Publishing Limited',
4      'Birmingham',
5      'UK',
6      'packtpub.com'
7  )
```

Listing 5-6. Inserting one row of data

To run this command, first open a new SQL editor window. Right-click on the database connection icon and select **Execute Command...**. Enter the SQL code as shown here. Then right-click anywhere in the window and select **Run Statement**.

To see that the data really has been inserted in to the table, right-click on the icon for the **PUBLISHERS** table (in the NetBeans **Services** tab) and select **View Data...**. An output window will appear:

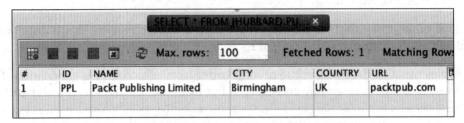

Figure 5-8. View of data in Publishers table

Notice the label at the top of this output window: **SELECT * FROM JHUBBARD.PU...**. This is the first part of the complete query that was executed to produce this output. You can see that complete query in line 1 of the SQL editor window above it:

```
SELECT * FROM JHUBBARD.PUBLISHERS FETCH FIRST 100 ROWS ONLY;
```

Figure 5-9. Query executed to produce output

There are two versions of the `insert into` statement. One version is illustrated in *Listing 5-6*. It requires a value for each field that is defined in the table. If any of those values is unknown or does not exist, then a more detailed version of the `insert into` statement is needed. This second version simply lists each field for which values are being provided.

The `insert` statement executed in *Listing 5-7* illustrates this second version:

```
insert into Publishers (id, name, city, country)
values ('A-V', 'Akademie-Verlag', 'Berlin', 'DE')
```

Listing 5-7. Insert specifying field names

This publisher has no URL, so that field must be left empty. Notice that the output in *Figure 5-10* shows that field value as **<NULL>**:

Figure 5-10. Current data in Publishers table

One advantage of this version is that you can list the fields in any order, as long as the order of their values matches.

Data can also be inserted into tables using `insert into` statements in **batch mode**. This simply means combining many of the statements together, one after the other, in a single SQL file, and then running the file itself, as we did with the `create table` commands in *Listing 5-5*.

Run this batch file in the SQL Editor:

```
1 insert into Publishers values ('PH', 'Prentice Hall, Inc.', 'Upper Saddle River, NJ', 'US', 'www.prenhall.com');
2 insert into Publishers values ('MHE', 'McGraw-Hill Education', 'New York, NY', 'US', 'www.mheducation.com');
3 insert into Publishers values ('A-W', 'Addison-Wesley Longman, Inc.', 'Reading, MA', 'US', 'www.awl.com');
4 insert into Publishers values ('CUP', 'Cambridge University Press', 'Cambridge', 'UK', 'www.cambridge.org');
```

Listing 5-8. Batch mode insertions

Then check the results:

#	ID	NAME	CITY	COUNTRY	URL
1	PPL	Packt Publishing Limited	Birmingham	UK	packtpub.com
2	A-V	Akademie-Verlag	Berlin	DE	<NULL>
3	PH	Prentice Hall, Inc.	Upper Saddle River, NJ	US	www.prenhall.com
4	MHE	McGraw-Hill Education	New York, NY	US	www.mheducation.com
5	A-W	Addison-Wesley Longman, Inc.	Reading, MA	US	www.awl.com
6	CUP	Cambridge University Press	Cambridge	UK	www.cambridge.org

Figure 5-11. Publishers table

The only problem with this approach is the tedium of writing all those `insert into` statements into the file. We shall see a more elegant solution later using JDBC instead.

Database queries

An SQL **database query** is an SQL statement that requests information from the database. If the query is successful, it results in a dataset that is usually presented in tabular form; that is, a virtual table. *Figure 5-9* shows an SQL query that was generated automatically when we clicked on **View Data...** in the drop down menu on the `Publishers` table. The key word to notice there is the first word, **select**.

The `select` statement has more clauses and variations than any other statement in SQL. It is fairly well covered on the Wikipedia page. We shall look at just a few variations in this section.

The query in *Listing 5-9* returns the name, city, and country of those publishers who are not US companies, sorting the results by name.

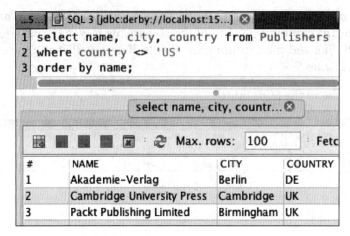

Listing 5-9. A query on the Publishers table

Both the where clause and the order by clause are optional. The symbol <> means "not equal".

The query in *Listing 5-10* shows that if you list two fields in an order by clause, the results will be sorted by the second field when tied in the first field.

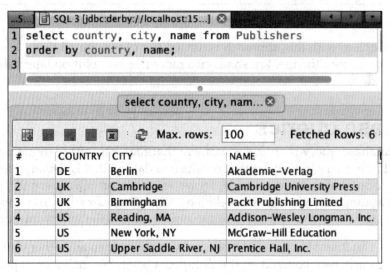

Listing 5-10. Query ordered by two fields

SQL data types

Standard SQL has four categories of data types: numbers, character strings, bit strings, and temporal types.

The numeric types include `int`, `smallInt`, `bigInt`, `double`, and `decimal(p,s)`, where `p` is the precision (the maximum number of digits) and `s` is the scale (the number of digits after the decimal point).

The character string types include `char(n)` and `varchar(n)`, as we have already seen.

The two bit string types are `bit(n)` and `bit varying(n)`. These can be interpreted as arrays of single bits—zeros and ones.

The temporal types include `date`, `time`, and `timestamp`. A timestamp is an instance in time that specifies the date and time, such as 2017-04-29 14:06. Notice that the ISO standard is used for dates; for example, April 29, 2017 is denoted as 2017-04-29.

Because of the competition among the major RDMS vendors, many SQL types have alternative names. For example, `integer` can be used in place of `int`, and `character` can be used in place of `char`. MySQL uses `int` and `char`.

JDBC

The **Java Database Connectivity (JDBC)** API is a library of Java packages and classes that provide easy execution of SQL commands on a RDB from within a Java program. This API is called a **database driver**. It can be downloaded from the provider of your DBMS, for example:

- Connector/J for MySQL
- `ojdbc6.jar` for Oracle Database 11g
- Microsoft JDBC Driver 6.0 for SQL Server

If you are using Java DB in NetBeans, no download is necessary; JDBC is already installed. But the library does have to be added to your project.

To use JDBC in a NetBeans project, follow these steps to add it to your project:

1. Right-click on the project icon (in the **Projects** tab) and select **Properties** from the drop-down menu.
2. In the **Project Properties** dialog, select **Libraries** under **Categories**.
3. In the **Compile** tab, select **Add Library**.

4. Then select **Java DB Driver** from the list of **Available Libraries** in the **Add Library** dialog, and click on the **Add Library** button.

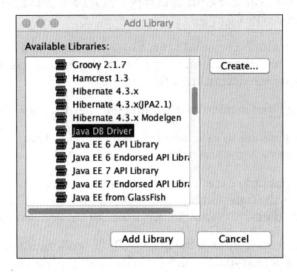

Figure 5-12. Adding the JDBC library in NetBeans

5. Then click **OK** in the **Project Properties** dialog.

To test your JDBC connection, run the simple test program, `JDBCTester`, which is shown in *Listing 5-11*:

```java
import java.sql.Connection;
import java.sql.DriverManager;
import java.sql.SQLException;
import java.sql.Statement;

public class JDBCTester {
    private static final String URL = "jdbc:derby://localhost:1527/Library";
    private static final String USR = "jhubbard";  // USE YOUR USERNAME HERE
    private static final String PWD = "dawj";      // USE YOUR PASSWORD HERE

    public static void main(String[] args) {
        try {
            Connection conn = DriverManager.getConnection(URL, USR, PWD);
            Statement stmt = conn.createStatement();

            stmt.close();
            conn.close();
        } catch (SQLException e) {
            System.err.println(e);
        }
    }
}
```

Listing 5-11. JDBC program access to a database

The `import` statements on lines 8-11 can be inserted automatically when you write lines 20-21 (just click on the tiny red ball on the line number where the import is needed, and then select the import statement to be inserted). Notice that these objects, `DriverManager`, `Connection`, `Statement`, and `SQLException` are all instances of classes that are defined in the `java.sql` package.

The three constants defined at lines 14-16 are used by the `DriverManager.getConnection()` method at line 20. The URL locates and identifies the database. The `USR` and `PWD` constants should be initialized with the username and password that you chose when you created the database. See *Figure 5-3*.

A `Statement` object is instantiated at line 21. It will contain an SQL statement that can be executed against your database. That code will fit in the area where the blank line 22 is now.

The Java program in *Listing 5-12* fills in the gap at line 22 of the program in *Listing 5-11*:

```java
     public static void main(String[] args) {
         try {
             Connection conn = DriverManager.getConnection(URL, USR, PWD);
             Statement stmt = conn.createStatement();

             String sql = String.format("select name, city from Publishers");
             ResultSet rs = stmt.executeQuery(sql);
             while (rs.next()) {
                 String pubName = rs.getString("name");
                 String pubCity = rs.getString("city");
                 System.out.printf("%s, %s%n", pubName, pubCity);
             }
             rs.close();

             stmt.close();
             conn.close();
         } catch (SQLException e) {
             System.err.println(e);
         }
     }
```

Listing 5-12. Program to print Publishers data

The SQL statement to be executed is defined as a `String` object at line 20. Line 23 passes that string to the `executeQuery()` method of our `stmt` object, and then saves the results in a `ResultSet` object.

Relational Databases

A `ResultSet` is a table-like object that provides more than 70 getter methods to access its contents. Since we asked for two fields (name and city) from the `Publishers` table, and we know that it has six rows, we can conclude that this `rs` object will act like a six by two array of `Strings`. The `while` loop at lines 26-30 will iterate down that array, one row at a time. At each row, we read the two fields separately, at lines 25-27, and then print them at line 28. The output is shown in *Figure 5-13*:

```
run:
Packt Publishing Limited, Birmingham
Prentice Hall, Inc., Upper Saddle River, NJ
McGraw-Hill Education, New York, NY
Addison-Wesley Longman, Inc., Reading, MA
Cambridge University Press, Cambridge
W. H. Freeman and Company, New York, NY
```

Figure 5-13. JDBC program output

Using a JDBC PreparedStatement

In the previous example, we used a JDBC **statement object** to query the `Library` database. We can also use the more flexible `PreparedStatement`, which includes variables that can be assigned different values dynamically.

The program in *Listing 5-13* shows how to insert new records into a database table. It reads the six lines of data from the CSV file shown in *Figure 5-14*, inserting the data in each line as a new record into the `Authors` table of our `Library` database:

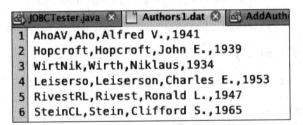

Figure 5-14. An external data file

To read from the file, we instantiate a `File` object at line 24 and a `Scanner` object at line 28.

At line 25, the SQL `insert` statement is coded into the `sql` string using four question marks (?) as place holders. These will be replaced by the actual data at lines 38-41. The `PreparedStatement` object, `ps`, is instantiated at line 28 using that `sql` string.

Each iteration of the `while` loop at lines 30-43 reads one line from the file. A separate scanner parses the line, reading the four CSV values at lines 33-37. Those four values are then pasted into the `PreparedStatement` at lines 37-40. On the third iteration, they would be the strings `"WirthNik"`, `"Wirth"`, `"Niklaus"`, and the integer 1934, forming the complete SQL statement:

```
insert into Authors values('WirthNik', 'Wirth', 'Niklaus', 1934)
```

This is then executed at line 41.

Since each execution of that `PreparedStatement` changes only one row in the table, the `rows` counter is incremented (by one) at line 41. So, the output from line 45 is:

```
6 rows inserted.
```

```java
public class AddAuthors {
    private static final String URL = "jdbc:derby://localhost:1527/Library";
    private static final String USR = "jhubbard";    // USE YOUR USERNAME HERE
    private static final String PWD = "dawj";        // USE YOUR PASSWORD HERE

    public static void main(String[] args) {
        try {
            Connection conn = DriverManager.getConnection(URL, USR, PWD);
            File file = new File("data/Authors.dat");
            String sql = "insert into Authors values(?, ?, ?, ?)";
            PreparedStatement ps = conn.prepareStatement(sql);
            Scanner fileScanner = new Scanner(file);
            int rows = 0;
            while (fileScanner.hasNext()) {
                String line = fileScanner.nextLine();
                Scanner lineScanner = new Scanner(line).useDelimiter(",");
                String id = lineScanner.next();
                String lastName = lineScanner.next();
                String firstName = lineScanner.next();
                int yob = lineScanner.nextInt();
                ps.setString(1, id);
                ps.setString(2, lastName);
                ps.setString(3, firstName);
                ps.setInt(4, yob);
                rows += ps.executeUpdate();
                lineScanner.close();
            }
            System.out.printf("%d rows inserted in Authors table.%n", rows);
            fileScanner.close();
            conn.close();
        } catch (IOException | SQLException e) {
            System.err.println(e);
        }
    }
}
```

Listing 5-13. JDBC program to insert data

Batch processing

The program in *Listing 5-13* suggests a general process, using JDBC `PreparedStatement` to load the complete database from external files. The following examples show how to do that:

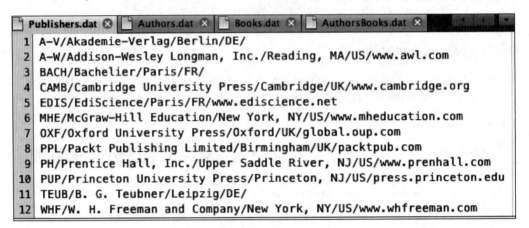

Figure 5-15. Data file for Publishers table

The data file shown in *Figure 5-15* contains data for 12 complete records for our `Publishers` table. The file has one record per line, separating the fields by the slash character (/). The program in the following screenshot loads all that data into the `Publishers` table:

```java
public class LoadPublishers {
    private static final String URL = "jdbc:derby://localhost:1527/Library";
    private static final String USR = "jhubbard";   // USE YOUR USERNAME HERE
    private static final String PWD = "dawj";       // USE YOUR PASSWORD HERE
    private static final String SQL =
            "insert into Publishers values(?, ?, ?, ?, ?)";
    private static final File DATA = new File("data/Publishers.dat");

    public static void main(String[] args) {
        try {
            Connection conn = DriverManager.getConnection(URL, USR, PWD);
            conn.createStatement().execute("delete from AuthorsBooks");
            conn.createStatement().execute("delete from Books");
            conn.createStatement().execute("delete from Publishers");
            PreparedStatement ps = conn.prepareStatement(SQL);
            Scanner fileScanner = new Scanner(DATA);
            int rows = 0;
            while (fileScanner.hasNext()) {
                String line = fileScanner.nextLine();
                Scanner lineScanner = new Scanner(line).useDelimiter("/");
                String id = lineScanner.next();
                String name = lineScanner.next();
                String city = lineScanner.next();
                String country = lineScanner.next();
                String url = (lineScanner.hasNext() ? lineScanner.next() : "");
                ps.setString(1, id);
                ps.setString(2, name);
                ps.setString(3, city);
                ps.setString(4, country);
                if (url.length() > 0) {
                    ps.setString(5, url);
                } else {
                    ps.setNull(5, Types.VARCHAR);
                }
                rows += ps.executeUpdate();
                lineScanner.close();
            }
            System.out.printf("%d rows inserted in Publishers table.%n", rows);
            conn.close();
        } catch (IOException | SQLException e) {
            System.err.println(e);
        }
    }
}
```

Listing 5-14. Program to load data into the Publishers table

The `PreparedStatement` is defined at line 24 as the constant string SQL, and the data file is defined as the `File` object DATA at line 23.

Relational Databases

At lines 31-33, we first use anonymous `Statement` objects to execute `delete from` statements on the `AuthorsBooks`, `Books`, and `Publishers` tables to empty them. We want the `Publishers` table to be emptied so that it will eventually contain only those records that are in this data file. And we must first empty the `AuthorsBooks` and `Books` tables because of their foreign key constraints (see *Figure 5-2*). SQL will not allow a record to be deleted if it is referenced by another existing record.

The rest of the code in *Listing 5-14* is like that in *Listing 5-13*, except for the processing of the `url` field. As we can see in lines 1 and 3 of the data file (*Figure 5-15*), some `Publishers` records do not have values for the `url` field (those publishers went out of business before the internet was invented). In situations like this, the database table should have the `NULL` value in those missing slots. In this case, we want the `NULL` value of the `varchar()` type. That is accomplished by the code at line 50:

```
ps.setNull(5, Types.VARCHAR);
```

We first use the conditional expression operator at line 44 to store the empty string in the `url` variable if the end-of-line has been reached in parsing the data file. Then the `if` statement at line 44 does the right thing, whether the `url` exists or not.

Data for the `Authors` table is shown in *Figure 5-16*:

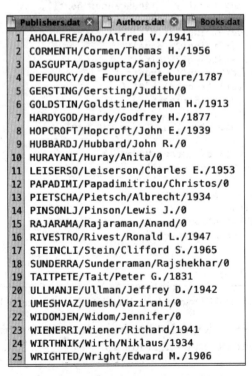

Figure 5-16. Data for the Authors table

These 25 records can be loaded into the `Authors` table using the program in *Listing 5-15*. It works the same way as the program in *Listing 5-14*.

The Packt website for this book has all these files and programs available for download. It also includes data files and programs for loading the other two tables (`Books` and `AuthorsBooks`) for the `Library` database.

```java
public class LoadAuthors {
    private static final String URL = "jdbc:derby://localhost:1527/Library";
    private static final String USR = "jhubbard";   // USE YOUR USERNAME HERE
    private static final String PWD = "dawj";       // USE YOUR PASSWORD HERE
    private static final String SQL = "insert into Authors values(?, ?, ?, ?)";
    private static final File DATA = new File("data/Authors.dat");

    public static void main(String[] args) {
        try {
            Connection conn = DriverManager.getConnection(URL, USR, PWD);
            conn.createStatement().execute("delete from AuthorsBooks");
            conn.createStatement().execute("delete from Authors");
            PreparedStatement ps = conn.prepareStatement(SQL);
            Scanner fileScanner = new Scanner(DATA);
            int rows = 0;
            while (fileScanner.hasNext()) {
                String line = fileScanner.nextLine();
                Scanner lineScanner = new Scanner(line).useDelimiter("/");
                String id = lineScanner.next();
                String lastName = lineScanner.next();
                String firstName = lineScanner.next();
                int yob = lineScanner.nextInt();
                ps.setString(1, id);
                ps.setString(2, lastName);
                ps.setString(3, firstName);
                ps.setInt(4, yob);
                rows += ps.executeUpdate();
                lineScanner.close();
            }
            System.out.printf("%d rows inserted in Authors table.%n", rows);
            conn.close();
        } catch (IOException | SQLException e) {
            System.err.println(e);
        }
    }
}
```

Listing 5-15. Program to load data into the Authors table

Database views

A database **view** is essentially a virtual table. It is created by embedding a `select` statement within a `create view` statement. You can then apply `select` queries to that view as though it were an actual table or result set.

The query shown in *Listing 5-16* lists those authors who have published books in the US. The `from` clause at line 2 lists all four tables because the query must link fields in the `Authors` table to fields in the `Publishers` table (see *Figure 5-2*). The three conditions at lines 3-5 correspond to the three foreign keys in the database. The `order by` clause at line 7 sorts the results alphabetically by last name, and the keyword `distinct` at line 1 removes duplicate rows from the output.

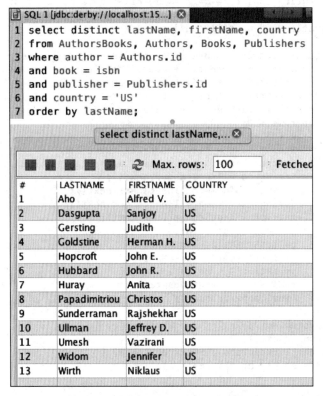

Listing 5-16. Selecting American authors

The output from this (or any) query is a **virtual table**. It has the same structure as an ordinary database table, but its data is not stored here as a separate table. A virtual table stores only references to real data that is stored in real tables elsewhere. Nevertheless, the concept of the virtual table is so useful that SQL allows us to name it and then use it as we would a real table. This is called a database **view**.

The view named AmericanAuthors is created in *Listing 5-17*. It simply prefaces the previous select statement with the code create view AmericanAuthors as in line 1.

```
create view AmericanAuthors as
    select distinct lastName, firstName, country
    from AuthorsBooks, Authors, Books, Publishers
    where author = Authors.id
    and book = isbn
    and publisher = Publishers.id
    and country = 'US'
    order by lastName;
```

Listing 5-17. Creating the View AmericanAuthors

You can see the resulting view object listing among the other database objects in the NetBeans **Services** tab window.

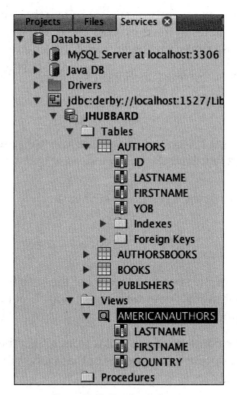

Figure 5-17. Database objects

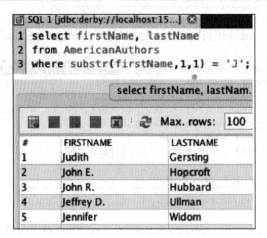

Listing 5-18. Querying the AmericanAuthors view

Now you can select from the AmericanAuthors view the same way you would from any of the four real tables in the database. The query in *Figure 5-18* finds all the authors in that view whose first name begins with the letter J.

Notice the use of the SQL substring function in this query:

substr(firstName,1,1)

 Some RDBMS vendors, notably Microsoft SQL Server, name the substring function substring instead of substr.

This returns the first letter of the firstName string.

In general, the call

substr(string, n, length)

will return the substring of the specified string of the specified length, beginning with the nth character. But in SQL, counting begins with 1, not 0 as in Java, so the first character is character number 1.

Views are dynamic. If the contents of the tables from which the view was created change, then subsequent output from queries of that view will change accordingly.

For example, suppose we correct the first name of the author GERSTING from Judith to Judith L. That is done with the SQL update command, as shown in *Listing 5-19*. Then, if we query our AmericanAuthors view again, we see the change there.

```
update Authors
set firstName = 'Judith L.'
where id = 'GERSTING';
```

Listing 5-19. Changing the Data

A view is a virtual table—it looks like a table, but it has no data itself. You can run select statements on it, but you cannot run update statements on a view. To change the data, you must update the table where the data resides.

```
select firstName, lastName
from AmericanAuthors
where substr(firstName,1,1) = 'J';
```

#	FIRSTNAME	LASTNAME
1	Judith L.	Gersting
2	John E.	Hopcroft
3	John R.	Hubbard
4	Jeffrey D.	Ullman
5	Jennifer	Widom

Listing 5-20. Repeat query after changing data

Subqueries

A view is a database object that acts like a virtual table for queries. A **subquery** is essentially a temporary view that is defined within a query and exists only for that query. So, you can think of a subquery as a temporary virtual table.

The statement in *Listing 5-21* creates a view named EuropeanBooks that lists the isbn codes of all the books that were published outside of the US.

```
create view EuropeanBooks as
    select isbn
    from Books, Publishers
    where publisher = id
    and country <> 'US';
```

Listing 5-21. Creating EuropeanBooks View

The statement in *Listing 5-22* contains a subquery at lines 4-5 that will return the virtual table of isbns of all those European books. Then the main select statement, beginning on line 1, returns the authors of those books.

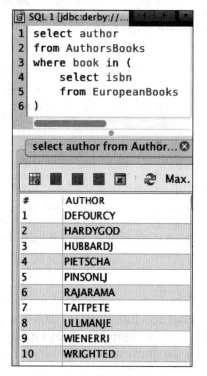

Listing 5-22. Authors of European books

Note that it is not necessary to use a view in the subquery. But it does make the query easier to understand. Like using a method in a Java program, the name of the method itself helps the code clarify what it is doing.

The query in *Listing 5-23* illustrates two SQL features. It uses the **aggregate function** `avg()`, which returns the average of the value of the specified field, `numPages` in this case. The query also specifies a **label** for the output: `Average`, which appears (in all capital letters) at the top of that output column. The output is the single value `447`.

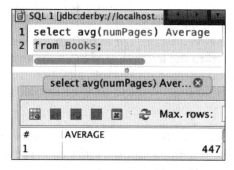

Listing 5-23. Average number of pages

There are five aggregate functions in standard SQL: `avg()`, `sum()`, `count()`, `max()`, and `min()`.

The query in *Listing 5-24* uses the previous code as a subquery, listing those books whose `numPages` value is above average. It also illustrates two other SQL features. It shows how a clause can follow a subquery; in this case, the `order by` clause. It also shows how to sort the output in descending order using the `desc` keyword.

```
select title, numPages
from Books
where numPages > (
    select avg(numPages)
    from Books
)
order by numPages desc
```

#	TITLE	NUMPAGES
1	Mathematical Structures for Computer Science	807
2	The Design and Analysis of Computer Algorithms	740
3	Data Structures with Java	613
4	A First Course in Database Systems	565
5	Oracle 10g Programming: A Primer	525
6	Lecons de Geometrie Analytique	491
7	Fundamentals of OOP and Data Structures in Java	463

Listing 5-24. Books with above average numbers of pages

Table indexes

A database table **index** is an object that is bound to a database table that facilitates searching that table.

Although the actual implementation of an index is usually proprietary and designed by the RDBMS vendor, it can be thought of as a separate file containing a multiway search tree that can locate keyed records in **logarithmic time**. This means that its lookup time is proportional to the logarithm of the number of records in the table. For example, if a lookup takes three probes into a table of 10,000 records, then it should be able to do a similar lookup into a table of 100,000,000 records in six probes (because $log\ n^2 = 2\ log\ n$).

Indexes take up extra space, and they take a bit of time to be updated when the table is changed by insertions or deletions. But if a large table is used mostly for lookups, then the database administrator is strongly advised to create an index for it.

A database index is usually implemented as a **B-tree** (invented by Rudolf Bayer and Ed McCreight when working at Boeing Research Labs in 1971), which stores key values and records addresses. The following figure illustrates this data structure:

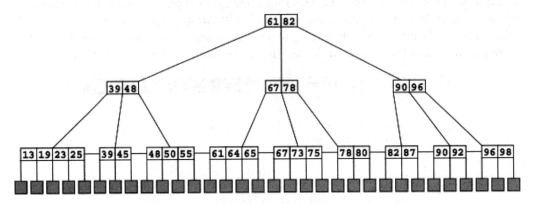

Figure 5-18. B-tree data structure

This could be an index on a table whose key is a two-digit positive integer. To find the record with key value 94, for example, the search begins at the root of the tree, which is the node at the top. That node contains 61 and 82. Since the target value is greater than those values, the search continues down the link on the right of 82, to the node containing 90 and 96. The target is between those two values, so the search continues down the link that is between them, to the next node. That node contains 90 and 92, so the search follows the link on the right of 92 and finally reaches the leaf node that contains the address of the record with the target key value, 94.

The B-tree in *Figure 5-18* has 13 key nodes containing a total of 31 keys at three levels, each node containing two four keys. But B-tree nodes can be configured to contain more than four nodes each. In fact, nodes with 20-40 keys each are not unusual. A B-tree with four levels of 40-key nodes could index a table holding 100,000,000,000 records. And yet, having only four levels means only four probes per search—that's nearly instantaneous!

The Java DB database system that comes with the NetBeans IDE automatically indexes each table that is created with a primary key. As with every object in a database, it has a name. As you can see in *Figure 5-19*, this index is named **SQL170430152720920** in the author's version of the database:

Figure 5-19. Default index on Authors table

If you are using an RDBMS that does not automatically index its keyed tables, then you can create an index with the SQL `create index` command:

```
create index AuthorsIndex on Authors(id)
```

This will create an index named `AuthorsIndex` on the `id` field of the `Authors` table.

Summary

In this chapter, we have described some of the basic features of relational databases and their access with Java programs: relational database design, primary and foreign keys, how to the SQL query language to create and access your databases, the JDBC **application programming interface** (**API**) for Java programming, database views, subqueries, and table indexes.

By working through the examples and running variations of the code provided here, you should have a comfortable understanding of how relational databases work with Java access.

6
Regression Analysis

One of the fundamental purposes of data analysis is the **extrapolation** of data. That is, the process of discovering patterns in the data and then projecting those patterns to anticipate unobserved or even future behavior. When a dataset appears to follow a pattern that looks like a mathematical function, then the algorithms used to identify that function or class of functions are characterized as **regression analysis**. In the simplest cases, when those functions are linear functions, the analysis is called **linear regression**.

Linear regression

The term "regression" was coined by the English statistician Francis Galton, who also originated the concept of correlation. Galton pioneered the field of data analysis with his studies of heredity. One of those early studies was on the heights of fathers and their sons, in which he observed that the sons of tall fathers tended toward more average heights. The title of that famous research paper was *Regression towards Mediocrity in Hereditary Stature*.

Figure 6.1: Sir Francis Galton

Regression Analysis

Linear regression is the simplest form of general regression analysis. The main idea here is to find numbers m and b, so that the line whose equation is $y = mx + b$ will fit closely through the given dataset of (x, y) points. The constants m and b are the slope and the y intercept.

Linear regression in Excel

Microsoft Excel is good at doing regression analysis. *Figure 6.2* shows an example of linear regression in Excel. The dataset is shown at the upper left corner in columns A-B, lines 1-11. There are two variables, Water and Dextrose, and 10 data points. The data comes from an experiment (source: J. L. Torgesen, V. E. Brown, and E. R. Smith, *Boiling Points of Aqueous Solutions of Dextrose within the Pressure Range of 200 to 1500 Millimeters*, J. Res. Nat. Bur. Stds. 45, 458-462 (1950).) that measured the boiling points of water and a dextrose solution at various pressures. It suggests a correlation between the two boiling points; that is, that they are affected by pressure in the same way. Although we will let x be the water boiling points and y be the dextrose solution boiling points, we are not suggesting that y depends upon x. In fact, they both depend upon the unmeasured variable, pressure. But, consequently, we do expect to see a linear relationship between these choices for x and y.

Here are the steps we followed to produce the results shown in *Figure 6.2*:

1. Enter the dataset, as shown in the upper left corner of the spreadsheet.
2. Select **Data Analysis** from the **Data** tab in the **Toolbar**.
3. Select **Regression** from the pop-up menu.
4. Fill in the **Input** and **Output** fields as shown in *Figure 6.3*.
5. Click on **OK**.

The results should look like *Figure 6.2*, except without the plot.

To run regression and most other statistical analysis tools in Excel, you must first activate its Analysis ToolPak. To do that, select **File | Options | AddIns | Analysis ToolPak** and then click on **OK**.

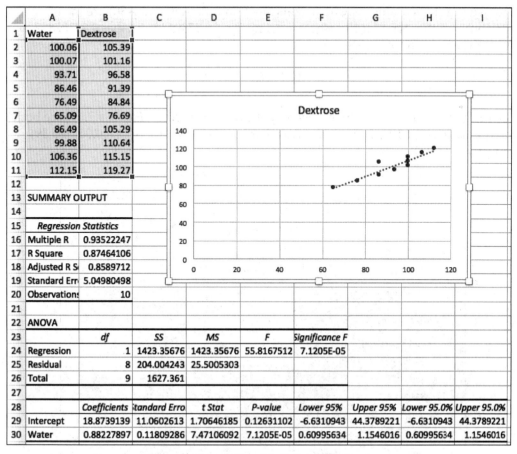

	A	B	C	D	E	F	G	H	I
1	Water	Dextrose							
2	100.06	105.39							
3	100.07	101.16							
4	93.71	96.58							
5	86.46	91.39							
6	76.49	84.84							
7	65.09	76.69							
8	86.49	105.29							
9	99.88	110.64							
10	106.36	115.15							
11	112.15	119.27							
12									
13	SUMMARY OUTPUT								
14									
15	Regression Statistics								
16	Multiple R	0.93522247							
17	R Square	0.87464106							
18	Adjusted R S	0.8589712							
19	Standard Err	5.04980498							
20	Observations	10							
21									
22	ANOVA								
23		df	SS	MS	F	Significance F			
24	Regression	1	1423.35676	1423.35676	55.8167512	7.1205E-05			
25	Residual	8	204.004243	25.5005303					
26	Total	9	1627.361						
27									
28		Coefficients	Standard Erro	t Stat	P-value	Lower 95%	Upper 95%	Lower 95.0%	Upper 95.0%
29	Intercept	18.8739139	11.0602613	1.70646185	0.12631102	-6.6310943	44.3789221	-6.6310943	44.3789221
30	Water	0.88227897	0.11809286	7.47106092	7.1205E-05	0.60995634	1.1546016	0.60995634	1.1546016

Figure 6.2: Linear Regression in Microsoft Excel

To obtain the plot, follow these additional steps:

1. Select the 22 cells, **A1-B11**.
2. Click on the X Y (Scatter) icon on the **Insert** tab, and then select the **Scatter** button (the first one) on the pop-up graphic menu.
3. In the resulting scatter plot, click on one of the points to select them all. Then right-click on one of the selected points and select **Add Trendline...** from the pop-up menu.

The dotted line on the scatter plot is the regression line for that data. (Excel calls it a "trendline.") It is the line that best fits the data points. Among all possible lines, it is the one that minimizes the sum of the squares of the vertical distances to the data points.

Regression Analysis

The two numbers under the Coefficients label (in cells **A29** and **A30**) are the y intercept and the slope for the regression line: $b = 18.8739139$ and $m = 0.88227897$. Thus, the equation for that line is:

$$y = 0.88228x + 18.874$$

The image in *Figure 6.4* is a magnification of the scatter plot in *Figure 6.2*, where both x and y are around 100. The three points there are (100.07, 101.16), (100.06, 105.39), and (99.88, 110.64).

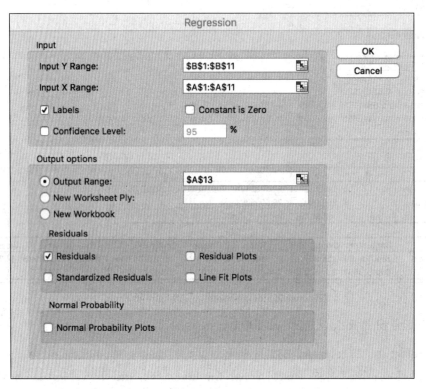

Figure 6.3: Microsoft Excel Linear Regression Dialog

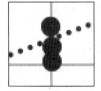

Figure 6.4: Magnification

The second point is the closest to the regression line. The corresponding point on the regression line that has the same x value is (100.06, 107.15). So, the vertical distance from that second point to the line is 107.15 − 105.39 = 1.76. This distance is called the **residual** for that data point. It is the sum of the squares of these residuals that is minimized by the regression line. (Why squares? Because of the Pythagorean Theorem. Distances in Euclidean space (where we live) are computed by summing the squares of the differences of the points' coordinates.)

In addition to the y intercept, the b slope, and m for the regression line, the Excel generates 29 other statistics. The first one, labelled Multiple R, under regression statistics, is the **sample correlation coefficient**:

$$r = 0.93522247$$

It can be computed from the formula:

$$r = \frac{n\Sigma xy - (\Sigma x)(\Sigma y)}{\sqrt{n\Sigma x^2 - (\Sigma x)^2}\sqrt{n\Sigma y^2 - (\Sigma y)^2}}$$

To see how this formula works, let's calculate r for an artificially simple dataset. The dataset in *Table 6.1* contains three data points: (1, 4), (2, 5), and (3, 7).

x	y	x^2	y^2	xy
1	4	1	16	4
2	5	4	25	10
3	7	9	49	21
6	16	14	90	35

Table 6.1: Computing r

The numbers in the 14 shaded cells are all computed from those six given coordinates. Each number in the bottom row is the sum of the three numbers above it. For example:

$$\Sigma y^2 = y_1^2 + y_2^2 + y_3^2 = 4^2 + 5^2 + 7^2 = 16 + 25 + 49 = 90$$

The number of data points is $n = 3$. Thus:

$$r = \frac{3(35) - (6)(16)}{\sqrt{3(14) - (6)^2}\sqrt{3(90) - (16)^2}} = \frac{105 - 96}{\sqrt{6}\sqrt{14}} = 0.982$$

Regression Analysis

The regression line for these three data points is shown in *Figure 6.5*:

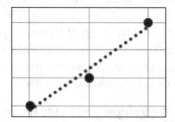

Figure 6.5: Reg. Line for Three Points

Clearly, the points are very close to the line. That is reflected in the value of the correlation coefficient, $r = 0.982$. That can be interpreted as saying that y is 98.2% linearly correlated with x.

To see what 100% linear correlation would mean, move the third point down, from (3, 7) to (3, 6), and then re-compute r:

$$r = \frac{3(32)-(6)(15)}{\sqrt{3(14)-(6)^2}\sqrt{3(77)-(15)^2}} = \frac{96-90}{\sqrt{6}\sqrt{6}} = \frac{6}{6} = 1.00$$

Those three points lie exactly on the line $y = x + 3$.

At the other extreme, move the third point down to (3, 4) and re-compute r again:

$$r = \frac{3(26)-(6)(13)}{\sqrt{3(14)-(6)^2}\sqrt{3(57)-(13)^2}} = \frac{76-76}{\sqrt{6}\sqrt{8}} = \frac{0}{4\sqrt{3}} = 0.00$$

This is a case of 0% correlation.

Computing the regression coefficients

The equation for the regression line is $y = mx + b$, where the constants m and b are computed from the coordinates of the points in the given dataset. Their formulas are:

$$m = \frac{n\sum xy - (\sum x)(\sum y)}{n(\sum x^2)-(\sum x)^2}$$

$$b = \frac{(\sum y)(\sum x^2)-(\sum x)(\sum xy)}{n(\sum x^2)-(\sum x)^2}$$

There are four sums here, each one computed directly from the data. We can simplify the formulas a little by renaming these sums A, B, C, and D:

$$A = \sum_{i=1}^{n} x_i$$
$$B = \sum_{i=1}^{n} y_i$$
$$C = \sum_{i=1}^{n} x_i^2$$
$$D = \sum_{i=1}^{n} x_i y_i$$

Then we have:

$$m = \frac{nD - AB}{nC - A^2}$$
$$b = \frac{BC - AD}{nC - A^2}$$

For example, given the preceding three-point dataset in *Table 6.1*, A = 6, B = 16, C = 14, and D = 35, so:

$$m = \frac{3(35) - (6)(16)}{3(14) - (6)^2} = \frac{9}{6} = 1.5$$

$$b = \frac{(16)(14) - (6)(35)}{3(14) - (6)^2} = \frac{14}{6} = 2.33$$

So, the equation for the regression line is y = 1.5x + 2.33.

The derivation of the formulas for *m* and *b* is derived here using multivariate calculus. To find the values of *m* and *b* that minimize this function:

$$f(m, b) = \sum_{i=1}^{n} (y_i - \hat{y}_i)^2 = \sum_{i=1}^{n} (y_i - mx_i - b)^2$$

Regression Analysis

We take its partial derivatives with respect to m and to b, set them both to zero, and then solve the resulting two equations simultaneously for m and b:

$$\frac{\partial}{\partial m}f(m,b) = \frac{\partial}{\partial m}\sum(y_i - mx_i - b)^2 = \sum 2(y_i - mx_i - b)\frac{\partial}{\partial m}(y_i - mx_i - b)$$

$$= \sum 2(y_i - mx_i - b)(-x_i) = -2\left[\sum x_i y_i - m\sum x_i^2 - b\sum x_i\right] = -2[D - mC - bA]$$

$$\frac{\partial}{\partial b}f(m,b) = \frac{\partial}{\partial b}\sum(y_i - mx_i - b)^2 = \sum 2(y_i - mx_i - b)\frac{\partial}{\partial b}(y_i - mx_i - b)$$

$$= \sum 2(y_i - mx_i - b)(-1) = -2\left[\sum y_i - m\sum x_i - b\sum 1\right] = -2[B - mA - bn]$$

Setting these two expressions to zero gives:

$$Cm + Ab = D$$
$$Am + nb = B$$

Or, equivalently:

$$\left(\sum x_i^2\right)m + \left(\sum x_i\right)b = \sum x_i y_i$$
$$\left(\sum x_i\right)m + nb = \sum y_i$$

These two equations are called the **normal equations** for the regression line.

 [The term **normal** here is not the same use as in the normal distribution.]

We can now solve the two normal equations simultaneously (for example, using Cramer's Rule) to obtain the formulas for m and b:

$$m = \frac{nD - AB}{nC - A^2}$$

$$b = \frac{BC - AD}{nC - A^2}$$

Finally, if we divide both sides of the second normal equation by n, we obtain:

$$\left(\frac{1}{n}\sum x_i\right)m + b = \frac{1}{n}\sum y_i$$

This is, more simply,

$$\bar{x}m + b = \bar{y}$$

In other words, the point whose coordinates are the means $\bar{x}$ and $\bar{y}$ lies on the regression line. That fact gives us a simpler formula for computing b in terms of m, assuming that the means have already been computed:

$$b = \bar{y} - \bar{x}m$$

Variation statistics

To clarify the meaning of the correlation coefficient r, let's first recall the sample variance that we looked at in Chapter 4, Statistics – Elementary Statistical Methods and Their Implementation in Java. For the y coordinates $\{y_1, y_2, \ldots, y_n\}$ of a dataset, the formula is:

$$s^2 = \frac{1}{n}\sum_{i=1}^{n}(y_i - \bar{y})^2$$

This is simply the average of the squared deviations of the values from their mean. For example, for the set $\{4, 7, 6, 8, 5\}$, the mean is $\bar{y} = 6$, so:

$$s^2 = \frac{1}{5}\left[(4-6)^2 + (7-6)^2 + (6-6)^2 + (8-6)^2 + (5-6)^2\right] = 2$$

If we compute the sum without then dividing by n, we have what's called the **total variation**:

$$TV = \sum_{i=1}^{n}(y_i - \bar{y})^2$$

In this little example, $TV = 10$. That's a simple measure of how much the data varies from its mean.

Regression Analysis

With linear regression, we can use the equation of the regression line, y = mx + b, to compute the y values of the points on that line that have the same x coordinates that are in the dataset. These y values are denoted by $\hat{y}_i$. So, for each i:

$$\hat{y}_i = mx_i + b$$

In our Water-Dextrose example in *Figure 6.2*, the eighth data point (on line 9 in *Figure 6.2*) is:

$$(x_8, y_8) = (99.88, 110.64)$$

So, $x_8 = 99.88$ and $y_8 = 110.64$. The regression line is:

$$y = 0.88228x + 18.874$$

So $\hat{y}_8$ is:

$$\hat{y}_8 = 0.88228x_8 + 18.874 = 0.88228(99.88) + 18.874 = 106.00$$

This y value is called an **estimate of y on x**. This is, of course, different from the mean value $\bar{y} = 100.64$.

The actual y value for x_8 is $y_8 = 110.64$. So, we can look at three differences:

The deviation of y_8 from its mean $\bar{y}$:

$$y_8 - \bar{y} = 110.64 - 100.64 = 10.00$$

The deviation of y_8 from its estimate (also called its residual) $\hat{y}_8$:

$$y_8 - \hat{y}_8 = 110.64 - 106.00 = 4.64$$

And the deviation the estimate $\hat{y}_8$ from mean $\bar{y}$:

$$\hat{y}_8 - \bar{y} = 106.00 - 100.64 = 5.36$$

Obviously, the first difference is the sum of the other two:

$$(y_8 - \bar{y}) = (y_8 - \hat{y}_8) + (\hat{y}_8 - \bar{y})$$

So, in general, we think of the deviation $(y_i - \bar{y})$ as the sum of two parts: the residual $(y_i - \hat{y}_i)$ and the factor $(\hat{y}_i - \bar{y})$.

As we saw in the previous section, it is a mathematical fact that the mean point $(\bar{x}, \bar{y})$ always lies on the regression line. Since the $\hat{y}_i$-values are computed from the equation of the regression line, we can think of that factor $(\hat{y}_i - \bar{y})$ as being "explained" by those regression line calculations. So, the deviation $(y_i - \bar{y})$ is the sum of two parts: the unexplained residual part $(y_i - \hat{y}_i)$, and the explained factor $(\hat{y}_i - \bar{y})$:

$$(y_i - \bar{y}) = (y_i - \hat{y}_i) + (\hat{y}_i - \bar{y})$$

Squaring both sides gives:

$$(y_i - \bar{y})^2 = (y_i - \hat{y}_i)^2 + (\hat{y}_i - \bar{y})^2 + 2(y_i - \hat{y}_i)(\hat{y}_i - \bar{y})$$

Then summing:

$$\sum_{i=1}^{n}(y_i - \bar{y})^2 = \sum_{i=1}^{n}(y_i - \hat{y}_i)^2 + \sum_{i=1}^{n}(\hat{y}_i - \bar{y})^2 + 2\sum_{i=1}^{n}(y_i - \hat{y}_i)(\hat{y}_i - \bar{y})$$

But that last sum is zero:

$$\sum(y_i - \hat{y}_i)(\hat{y}_i - \bar{y}) = \sum(y_i - mx_i - b)(mx_i + b - \bar{y})$$
$$= m\sum x_i(y_i - mx_i - b) + (b - \bar{y})\sum(y_i - mx_i - b)$$
$$= m(0) + (b - \bar{y})(0) = 0$$

This is because:

$$\sum x_i(y_i - mx_i - b) = \sum x_i y_i - \left(\sum x_i^2\right)m - \left(\sum x_i\right)b = 0$$
$$\sum(y_i - mx_i - b) = \sum y_i - \left(\sum x_i\right)m - nb = 0$$

These equation are from the Normal Equations (in the previous section).

Thus:

$$\sum_{i=1}^{n}(y_i - \bar{y})^2 = \sum_{i=1}^{n}(y_i - \hat{y}_i)^2 + \sum_{i=1}^{n}(\hat{y}_i - \bar{y})^2$$

Or:

$$TV = UV + EV$$

Here,

$$TV = \sum_{i=1}^{n}(y_i - \bar{y})^2$$
$$UV = \sum_{i=1}^{n}(y_i - \hat{y}_i)^2$$
$$EV = \sum_{i=1}^{n}(\hat{y}_i - \bar{y})^2$$

These equations define the **total variation** TV, the **unexplained variation** UV, and the **explained variation** EV.

The variation data in the Water-Dextrose example from *Figure 6.2* is reproduced here in *Figure 6.6*.

22	ANOVA		
23		df	SS
24	Regression	1	1423.35676
25	Residual	8	204.004243
26	Total	9	1627.361

Figure 6.6: Variation in W-D Experiment

You can read these values under the SS column (for "sum of squares") of the ANOVA (for "analysis of variance") section:

$$TV = \sum_{i=1}^{n}(y_i - \bar{y})^2 = 1627.361$$

$$UV = \sum_{i=1}^{n}(y_i - \hat{y}_i)^2 = 204.0042$$

$$EV = \sum_{i=1}^{n}(\hat{y}_i - \bar{y})^2 = 1423.35676$$

Java implementation of linear regression

The Apache Commons Math library includes the package stat.regression. That package has a class named SimpleRegression which includes methods that return many of the statistics discussed in this chapter. (See *Appendix, Java Tools* on how to install this library in NetBeans.)

The data file Data1.dat, shown in *Figure 6.7*, contains the data from the Water-Dextrose experiment presented earlier:

```
Data1.dat   Example1.java
 1  Boiling Temperatures
 2  10
 3  Water      Dextrose Solution
 4  100.06     105.39
 5  100.07     101.16
 6   93.71      96.58
 7   86.46      91.39
 8   76.49      84.84
 9   65.09      76.69
10   86.49     105.29
11   99.88     110.64
12  106.36     115.15
13  112.15     119.27
```

Figure 6.7: Data Source for Example1

Regression Analysis

The data values on lines 4-13 are separated by tab characters. The program in *Listing 6.1* reads that data, uses a `SimpleRegression` object to extract and then prints the statistics shown in the output panel. These are the same results that we obtained earlier for this data.

```java
import org.apache.commons.math3.stat.regression.SimpleRegression;

public class Example1 {
    public static void main(String[] args) {
        SimpleRegression sr = getData("data/Data1.dat");
        double m = sr.getSlope();
        double b = sr.getIntercept();
        double r = sr.getR();   // correlation coefficient
        double r2 = sr.getRSquare();
        double sse = sr.getSumSquaredErrors();
        double tss = sr.getTotalSumSquares();

        System.out.printf("y = %.6fx + %.4f%n", m, b);
        System.out.printf("r = %.6f%n", r);
        System.out.printf("r2 = %.6f%n", r2);
        System.out.printf("EV = %.5f%n", tss - sse);
        System.out.printf("UV = %.4f%n", sse);
        System.out.printf("TV = %.3f%n", tss);
    }
}
```

```
run:
y = 0.882279x + 18.8739
r = 0.935222
r2 = 0.874641
EV = 1423.35676
UV = 204.0042
TV = 1627.361
```

Listing 6.1: Using a SimpleRegression Object

At *line 20*, the variable `sse` is assigned the value returned by the method `sr.getSumSquaredErrors()`. This sum of squared errors is the unexplained variation (UV), which Excel calls the Residual. The output shows its value is `204.0042`, which agrees with *Figure 6.6*.

At *line 21*, the variable `tss` is assigned the value returned by the method `sr.getTotalSumSquares()`. This "total sum of squares" is the **total variation** (**TV**). The output shows that its value is 1627.361, also agreeing with *Figure 6.6*.

The `import` statement at *line 11* shows where the `SimpleRegression` class is defined. It is instantiated at *line 32* in the `getData()` method shown in *Listing 6.2*.

```
 Data1.dat  Example1.java
31      public static SimpleRegression getData(String data) {
32          SimpleRegression sr = new SimpleRegression();
33          try {
34              Scanner fileScanner = new Scanner(new File(data));
35              fileScanner.nextLine();   // read past title line
36              int n = fileScanner.nextInt();
37              fileScanner.nextLine();   // read past line of labels
38              fileScanner.nextLine();   // read past line of labels
39              for (int i = 0; i < n; i++) {
40                  String line = fileScanner.nextLine();
41                  Scanner lineScanner = new Scanner(line).useDelimiter("\\t");
42                  double x = lineScanner.nextDouble();
43                  double y = lineScanner.nextDouble();
44                  sr.addData(x, y);
45              }
46          } catch (FileNotFoundException e) {
47              System.err.println(e);
48          }
49          return sr;
50      }
51  }
```

Listing 6.2: The getData() Method from Example1

The `fileScanner` object instantiated at *line 34* reads lines from the data file that is located by the `data` string passed to it. At *lines 35-38*, it reads the first three lines of the file, saving the number of data points in the variable n. Then the `for` loop at *lines 39-45* reads each of those *n* data lines, extracts the *x* and *y* values, and then adds them to the `sr` object at *line 44*. Note that the `lineScanner`, instantiated at line 41, is told to use the tab character `'\t'` as its delimiter. (The backslash symbol \ has to be "escaped" with a preceding backslash for Java to recognize it as such. The expression `\\t`, then, is a string that contains that single tab character.)

Regression Analysis

The program Example2, shown in *Listing 6.3*, is similar to the program Example1. They have the same output:

```java
public class Example2 {
    private static double sX=0, sXX=0, sY=0, sYY=0, sXY=0;
    private static int n=0;

    public static void main(String[] args) {
        getData("data/Data1.dat");
        double m = (n*sXY - sX*sY)/(n*sXX - sX*sX);
        double b = sY/n - m*sX/n;
        double r2 = m*m*(n*sXX - sX*sX)/(n*sYY - sY*sY);
        double r = Math.sqrt(r2);
        double tv = sYY - sY*sY/n;
        double mX = sX/n;   // mean value of x
        double ev = (sXX - 2*mX*sX + n*mX*mX)*m*m;
        double uv = tv - ev;

        System.out.printf("y = %.6fx + %.4f%n", m, b);
        System.out.printf("r = %.6f%n", r);
        System.out.printf("r2 = %.6f%n", r2);
        System.out.printf("EV = %.5f%n", ev);
        System.out.printf("UV = %.4f%n", uv);
        System.out.printf("TV = %.3f%n", tv);
    }
}
```

```
run:
y = 0.882279x + 18.8739
r = 0.935222
r2 = 0.874641
EV = 1423.35676
UV = 204.0042
TV = 1627.361
```

Listing 6.3: Computing Statistics Directly

The difference is that `Example2` computes the statistics directly instead of using a `SimpleRegression` object.

The eight statistics are all computed at *lines 18-25* from the formulas developed in the previous sections. They are the slope m and y intercept b for the regression line, the correlation coefficient r and its square r^2, the mean $\bar{x}$, and the three variance statistics, *TV*, *EV*, and *UV*. These are computed from the five sums declared as global variables at *line 13*: $\sum x_i$, $\sum x_i^2$, $\sum y_i$, $\sum y_i^2$, and $\sum x_i y_i$.

The five sums, denoted as `sX`, `sXX`, `sY`, `sYY`, and `sXY`, are accumulated in the `for` loop at `lines 42-52`.

```
35      public static void getData(String data) {
36          try {
37              Scanner fileScanner = new Scanner(new File(data));
38              fileScanner.nextLine();  // read past title line
39              n = fileScanner.nextInt();
40              fileScanner.nextLine();  // read past line of labels
41              fileScanner.nextLine();  // read past line of labels
42              for (int i = 0; i < n; i++) {
43                  String line = fileScanner.nextLine();
44                  Scanner lineScanner = new Scanner(line).useDelimiter("\\t");
45                  double x = lineScanner.nextDouble();
46                  double y = lineScanner.nextDouble();
47                  sX += x;
48                  sXX += x*x;
49                  sY += y;
50                  sYY += y*y;
51                  sXY += x*y;
52              }
53          } catch (FileNotFoundException e) {
54              System.err.println(e);
55          }
56      }
57  }
```

Listing 6.4: The getData() Method from Example2

The program `Example3`, shown in *Listing 6.5*, encapsulates the code from `Example2` and generates the image in *Figure 6.8*.

```
11  public class Example3 {
12      public static void main(String[] args) {
13          Data data = new Data(new File("data/Data1.dat"));
14          JFrame frame = new JFrame(data.getTitle());
15          frame.setDefaultCloseOperation(JFrame.EXIT_ON_CLOSE);
16          RegressionPanel panel = new RegressionPanel(data);
17          frame.add(panel);
18          frame.pack();
19          frame.setSize(500, 422);
20          frame.setResizable(false);
21          frame.setLocationRelativeTo(null);   // center frame on screen
22          frame.setVisible(true);
23      }
24  }
```

Listing 6.5: Example3 Program

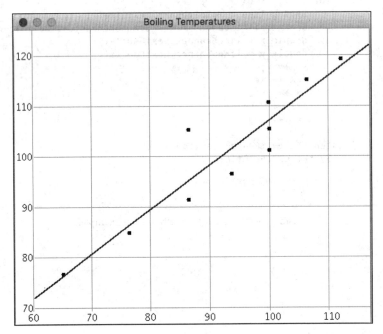

Figure 6-8. Example 3 Output

The program instantiates two auxiliary classes: a `Data` class at *line 13*, and a `RefressionPanel` class at *line 16*. The rest of the code here is setting up the `JFrame`.

The `Data` class constructor is shown in *Listing 6.6*. That's where the code from `Example2` is put.

```java
public class Data {
    private String title,xName, yName;
    private int n;
    private double[] x, y;
    private double sX, sXX, sY, sYY, sXY, minX, minY, maxX, maxY;
    private double meanX, meanY, slope, intercept, corrCoef;

    public Data(File inputFile) {
        try {
            Scanner input = new Scanner(inputFile);
            title = input.nextLine();
            n = input.nextInt();
            xName = input.next();
            yName = input.next();
            input.nextLine();
            x = new double[n];
            y = new double[n];
            minX = minY = Double.POSITIVE_INFINITY;
            maxX = maxY = Double.NEGATIVE_INFINITY;
            for (int i = 0; i < n; i++) {
                double xi = x[i] = input.nextDouble();
                double yi = y[i] = input.nextDouble();
                sX += xi;
                sXX += xi*xi;
                sY += yi;
                sYY += yi*yi;
                sXY += xi*yi;
                minX = (xi < minX? xi: minX);
                minY = (yi < minY? yi: minY);
                maxX = (xi > maxX? xi: maxX);
                maxY = (yi > maxY? yi: maxY);
            }
            meanX = sX/n;
            meanY = sY/n;
            slope = (n*sXY - sX*sY)/(n*sXX - sX*sX);
            intercept = meanY - slope*meanX;
            corrCoef = slope*Math.sqrt((n*sXX - sX*sX)/(n*sYY - sY*sY));
        } catch (FileNotFoundException e) {
            System.err.println(e);
        }
    }
}
```

Listing 6.6: The Data Class for Example 3

Regression Analysis

The rest of the `Data` class consists of 16 getter methods. They are listed in *Figure 6.9*:

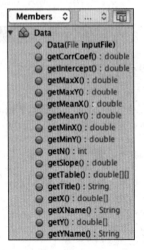

Figure 6.9: Data class members

The `RegressionPanel` class is shown in *Listing 6.7*:

```java
14  public class RegressionPanel extends JPanel {
        private static final int WIDTH=500, HEIGHT=400, BUFFER=28, MARGIN=40;
16      private final Data data;
        private double xMin, xMax, yMin, yMax, xRange, yRange, gWidth, gHeight;
        private double slope, intercept;
19
20      public RegressionPanel(Data data) {
21          this.data = data;
            this.setSize(WIDTH, HEIGHT);
23          this.xMin = data.getMinX();
24          this.xMax = data.getMaxX();
25          this.yMin = data.getMinY();
26          this.yMax = data.getMaxY();
27          this.slope = data.getSlope();
28          this.intercept = data.getIntercept();
29          this.xRange = xMax - xMin;
30          this.yRange = yMax - yMin;
31          this.gWidth = WIDTH - 2*MARGIN - BUFFER;
32          this.gHeight = HEIGHT - 2*MARGIN - BUFFER;
            setBackground(Color.WHITE);
34      }
35
36      @Override
        public void paintComponent(Graphics g) {
38          super.paintComponent(g);
39          Graphics2D g2 = (Graphics2D)g;
40          g2.setStroke(new BasicStroke(1));
41          drawGrid(g2);
42          drawPoints(g2, data.getX(), data.getY());
43          drawLine(g2);
44      }
```

Listing 6.7: RegressionPanel Class for Example 3

[162]

This contains all the graphics code (not shown here). The image is generated by the three methods called at *lines 41-43* in the `paintComponent()` method:

- `drawGrid()`
- `drawPoints()`
- `drawLine()`

The resulting image here can be compared to the Excel graph in *Figure 6.2*.

> All the code is available for download at the Packt website for this book.

Anscombe's quartet

The linear regression algorithm is appealing, because its implementation is straightforward and it is available from many different sources, such as Microsoft Excel or the Apache Commons Math Java API. However, its popularity sometimes leads the unsuspecting to misuse it.

The underlying assumption upon which the algorithm is based is that the two random variables, X and Y, actually are linearly related. The correlation coefficient r can help the researcher decide whether that assumption is valid for example,: yes if $r \approx \pm 1$, no if $r \approx 0$.

For example, the Water-Dextrose data has an r value of 93.5%, suggesting that the relationship is probably linear. But of course, certainty is not possible. The best the researcher can do to raise confidence would be to obtain more data and then "run the numbers" again. After all, a sample size of 10 is not very credible.

The English statistician Frank Anscombe suggested the four examples shown in *Figure 6.10*:

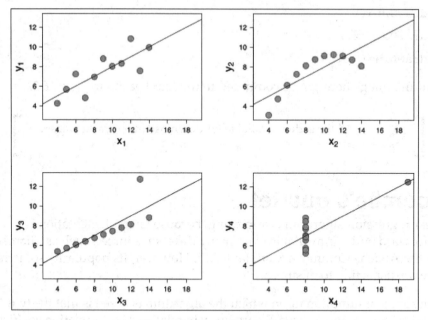

Figure 6.10: Anscombe's Quartet

This shows four 11-point datasets, to each of which the linear regression algorithm has been applied. The data points were chosen so that the same regression line results from each application. The four datasets also have the same mean, variance, and correlation.

Obviously, the data are very different. The scatterplot in the first example looks like the plot of the data points in the Water-Dextrose example. They had a linear correlation coefficient of 93.5%, so the regression line has some validity. In the second example, the data appears quadratic—like time-lapse photos of a projectile fired from the ground, such as a cannonball or a football. We might expect that correlation coefficient to be near 0. The third example has 10 points that are almost perfectly linear, with the 11th point way out of line. A good conclusion here might be that the outlier was an error in measurement or data entry.

The fourth example also has an **outlier** (the corner point). But this has a more severe problem: all the other points have the same x value. That will cause the algorithm itself to fail, because the denominator of the formula for the slope m will be nearly zero:

$$m = \frac{n\sum y_i^2 - \sum x_i \sum y_i}{n\sum x_i^2 - \left(\sum x_i\right)^2}$$

For example, if all the $x_i = 8$ (as they seem to here), then $\sum x_i = 80$ and $\sum x^2 = 640$, so the denominator would be $10(640) - (80)^2 = 0$.

The points are linear, but the line is vertical, so it cannot have an equation of the form $y = mx + b$.

To solve the problem revealed in the fourth example in Anscombe's quartet: just reverse the roles of x and y. This will result in an equation of the form $x = b_0 + b_1 y$, where the constants b_0 and b_1 are computed from the formulas:

$$b_1 = \frac{n\sum x_i y_i - \sum x_i \sum y_i}{n\sum y_i^2 - \left(\sum y_i\right)^2}$$

$$b_0 = \bar{x} - b_1 \bar{y}$$

For this particular dataset, it looks like the slope b_1 will be nearly zero.

Polynomial regression

The previous analysis has been centered around the idea of obtaining a linear equation to represent a given dataset. However, many datasets derive from non-linear relationships. Fortunately, there are alternative mathematical models from which to choose.

The simplest nonlinear functions are polynomials: $y = f(x) b_0 + b_1 x + b_2 x^2 + \ldots + b_d x^d$, where d is the degree of the polynomial and $b_0, b_1, b_2, \ldots, b_m$ are the coefficients to be determined.

Regression Analysis

Of course, a linear function is simply a first-degree polynomial: $y = b_0 + b_1 x$. We have already solved that problem (in the previous derivation, we called the coefficients m and b instead of b_1 and b_0). We used the **method of least squares** to derive the formulas for the coefficients:

$$b_1 = \frac{n\sum x_i y_i - \sum x_i \sum y_i}{n\sum x_i^2 - \left(\sum x_i\right)^2}$$

$$b_0 = \bar{y} - b_1 \bar{x}$$

Those formulas were derived from the normal equations:

$$\left(\sum x_i^2\right) b_1 + \left(\sum x_i\right) b_0 = \sum x_i y_i$$

$$\left(\sum x_i\right) b_1 + n b_0 = \sum y_i$$

The equations were obtained by minimizing the sum of squares:

$$\sum_{i=1}^{n} (y_i - \hat{y}_i)^2 = \sum_{i=1}^{n} (y_i - b_1 x_i - b_0)^2$$

We can apply the same least squares method to find the best-fitting polynomial of any degree d for a given dataset, provided that d is less than the number of independent points in the dataset.

For example, to find the best fitting polynomial of the degree $m = 2$, $f(x) = b_0 + b_1 x + b_2 x^2$ (also called the **least-squares parabola** for the given dataset), we would minimize
the sum:

$$\sum_{i=1}^{n} \left(y_i - b_2 x_i^2 - b_1 x_i - b_0\right)^2$$

This equation is used to determine the coefficients b_0, b_1, and b_2. From a calculus point of view, this expression has the form:

$$z = \sum \left(A_i + B_i u + C_i v + D_i w\right)^2$$

As before, we minimize z by setting its partial derivatives equal to zero:

$$\frac{\partial z}{\partial u} = \frac{\partial z}{\partial v} = \frac{\partial z}{\partial w} = 0$$

This results in the normal equations for the second-degree regression:

$$nb_0 + \left(\sum x_i\right)b_1 + \left(\sum x_i^2\right)b_2 = \sum y_i$$
$$\left(\sum x_i\right)b_0 + \left(\sum x_i^2\right)b_1 + \left(\sum x_i^3\right)b_2 = \sum x_i y_i$$
$$\left(\sum x_i^2\right)b_0 + \left(\sum x_i^3\right)b_1 + \left(\sum x_i^4\right)b_2 = \sum x_i^2 y_i$$

These three equations can be solved simultaneously for the three unknowns b_0, b_1, and b_2.

Here is a specific example that relates the stopping distance of a specific car to its speed at the moment that its brakes are applied. Our variables are:

- x is the speed of the car (in miles per hour) at the moment of braking
- y is the distance (in feet) that the car travels from the braking point to the stopping point

We assume that the brakes are applied uniformly during the stopping interval and that they are applied with the same force each time the experiment is performed.

Here is our dataset:

$$\{(20, 52), (30, 87), (40, 136), (50, 203), (60, 290), (70, 394)\}$$

It has $n = 6$ data points.

After computing all the sums for the normal equations, they are:

$$6b_0 + 270b_1 + 13{,}900b_2 = 1162$$
$$270b_0 + 13{,}900b_1 + 783{,}000b_2 = 64220$$
$$13{,}900b_0 + 783{,}000b_1 + 46{,}750{,}000b_2 = 3{,}798{,}800$$

Regression Analysis

These can be simplified to:

$$6b_0 + 270b_1 + 13{,}900b_2 = 1162$$
$$27b_0 + 1390b_1 + 78{,}300b_2 = 6422$$
$$139b_0 + 7830b_1 + 467{,}500b_2 = 37{,}988$$

There are several ways to solve a system of equations like this. You could use Cramer's Rule, which involves computing various determinants. Or you could use Gaussian Elimination, which systematically subtracts multiples of one row from another to eliminate all but a single variable in each row.

Instead, we will use the Apache Commons math library to solve these equations. The webpage explains how that works: http://commons.apache.org/proper/commons-math/userguide/linear.html.

The main part of the program is shown in *Listing 6.8*.

```java
import org.apache.commons.math3.linear.*;

public class Example4 {
    static int n = 6;
    static double[] x = {20, 30, 40, 50, 60, 70};
    static double[] y = {52, 87, 136, 203, 290, 394};
    static double[][] a;
    static double[] b;

    public static void main(String[] args) {
        double[][] a = new double[3][3];
        double[] w = new double[3];
        deriveNormalEquations(a, w);
        printNormalEquations(a, w);
        solveNormalEquations(a, w);
        printResults();
    }
```

```
run:
       6b0 +      270b1 +     13900b2 =      1162
     270b0 +    13900b1 +    783000b2 =     64220
   13900b0 +   783000b1 +  24010000b2 =   3798800
   f(t) = -115.67 + 6.950t + -0.00148t^2
   f(55) = 262.1
```

Listing 6-8. Main Program and Output for Example 4

The normal equations are encapsulated into the two arrays `a[][]` and `w[]`. The matrix equation $a \cdot b = w$ is equivalent to the system of three normal equations shown above.

The output shows the three normal equations, the resulting quadratic polynomial, and the computed value, $f(55) = 262.1$. This interpolation predicts that the stopping distance for the test car would be about 262 feet if the brakes were applied at 55 mph.

The derivation of the normal equations is shown in *Listing 6.9*. This directly implements the summation formulas shown earlier.

```java
    public static void deriveNormalEquations(double[][] a, double[] w) {
        int n = y.length;
        for (int i = 0; i < n; i++) {
            double xi = x[i];
            double yi = y[i];
            a[0][0] = n;
            a[0][1] = a[1][0] += xi;
            a[0][2] = a[1][1] = a[2][0] += xi*xi;
            a[1][2] = a[2][1] += xi*xi*xi;
            a[2][2] = xi*xi*xi*xi;
            w[0] += yi;
            w[1] += xi*yi;
            w[2] += xi*xi*yi;
        }
    }

    public static void printNormalEquations(double[][] a, double[] w) {
        for (int i = 0; i < 3; i++) {
            System.out.printf("%8.0fb0 + %6.0fb1 + %8.0fb2 = %7.0f%n",
                    a[i][0], a[i][1], a[i][2], w[i]);
        }
    }
```

Listing 6.9: Normal Equations for Example 4

The normal equations are solved in the method at lines 49-55, shown in *Figure 6.10*.

Regression Analysis

This program uses six classes from the `org.apache.commons.math3.linear` package.

```java
    Example4.java    Example5.java    Example6.java
49      private static double[] solveNormalEquations(double[][] a, double[] w) {
50          RealMatrix m = new Array2DRowRealMatrix(a, false);
51          LUDecomposition lud = new LUDecomposition(m);
52          DecompositionSolver solver = lud.getSolver();
53          RealVector v = new ArrayRealVector(w, false);
54          return solver.solve(v).toArray();
55      }
56
57      private static void printResults(double[] b) {
58          System.out.printf("f(t) = %.2f + %.3ft + %.5ft^2%n", b[0], b[1], b[2]);
59          System.out.printf("f(55) = %.1f%n", f(55, b));
60      }
61
62      private static double f(double t, double[] b) {
63          return b[0] + b[1]*t + b[2]*t*t;
64      }
65  }
```

Listing 6.10: Solving the Normal Equations

The code at line 50 instantiates a `RealMatrix` object m that represents the coefficient array `a[][]`. On the next line, an `LUDecomposition` object `lud` instantiates the **LU decomposition** of the matrix m into its lower and upper triangular factors (see next). Line 52 sets up a `solver` object that is then used at line 54 to solve the equations for the solution b. The `solve()` method requires a `RealVector` object, v, which we instantiate from the array `w[]` at line 53.

The instantiation of the `solver` object at line 52 performs an algorithm known as LU decomposition. The letters L and U stand for "lower" and "upper". The algorithm factors a given matrix M into two matrices, L and U, as $M = LU$, where L is a lower triangular matrix and U is an upper triangular matrix. This allows the solution of the matrix equation without having to invert the original matrix directly:

$$Mw = v$$
$$LUw = v$$
$$Uw = L^{-1}v$$
$$w = U^{-1}L^{-1}v$$

This method is significantly faster than Gaussian Elimination and generally much faster than Cramer's Rule.

The method of least squares, which we have been using to solve linear and quadratic regression problems, applies in the same way to general polynomials: $y = f(x) = b_0 + b_1x + b_2x^2+\ldots+b_dx^d$.

The only mathematical restriction for polynomial regression is that the degree of the polynomial be less than the number of data points (that is, $d < n$). This restriction should be evident in the simplest cases. For example, for linear regression ($d = 1$), two points determine a line. And for quadratic regression ($d = 2$), three points determine a parabola.

But the rule that "more data is better" almost always applies.

Multiple linear regression

The regression models that we have considered previously have all assumed a variable y depends upon only one independent variable x. Similar models work for multivariate functions that have any number k independent variables:

$$y = f(x_1, x_2, \ldots, x_k)$$

In this section, we will look at the regression model for **multilinear functions**:

$$y = a_0 + a_1x_1 + a_2x_2 + \cdots + a_kx_k$$

Here, we will develop the model for multilinear regression on $k = 2$ two independent variables:

$$z = a + bx + cy$$

We have renamed the variables and constants here to remind the reader of solid analytic geometry. This equation defines a plane in three-dimensional Euclidean space. The coordinates are x, y, and z. The plane intersects the z axis at $z = c$, where $x = y = 0$. The constants a and b are the slopes of the lines where the plane intersects the xz plane and yz plane.

The regression problem in this case is:

$$\left[\begin{array}{c} \text{ Given a dataset } \{(x_1, z_1), (x_2, z_2), \cdots, (x_n, z_n)\}, \text{ find constants } a, b, \text{ and } c \text{ that} \\ \text{minimize } \sum_{i=1}^{n}(z_i - \hat{z}_i)^2 = \sum_{i=1}^{n}(z_i - a - bx_i - cy_i)^2. \end{array} \right]$$

Regression Analysis

This is nearly the same as the quadratic regression problem that we solved:

> Given a dataset $\{(x_1, z_1), (x_2, z_2), \cdots, (x_n, z_n)\}$, find constants a, b, and c that minimize $\sum_{i=1}^{n}(z_i - \hat{z}_i)^2 = \sum_{i=1}^{n}(z_i - a - bx_i - cy_i^2)^2$.

(Here in the quadratic regression problem, we have renamed the dependent variable z to avoid confusion with the independent variable y in the multilinear problem.) The least squares algorithm is the same. In both cases, we will have three normal equations to solve for the three unknowns, a, b, and c.

The normal equations are:

$$na + \left(\sum x_i\right)b + \left(\sum y_i\right)c = \sum z_i$$
$$\left(\sum x_i\right)a + \left(\sum x_i^2\right)b + \left(\sum x_i y_i\right)c = \sum x_i z_i$$
$$\left(\sum y_i\right)a + \left(\sum x_i y_i\right)b + \left(\sum y_i^2\right)c = \sum y_i z_i$$

These can be derived just as before, applying partial derivatives to the sum of squares function. The sums can be tabulated by hand or machine.

Here is a dataset that suggests that a boy's weight z depends upon his age x and height y: {(10, 59, 71), (9, 57, 68), (12, 61, 76), (10, 52, 56), (9, 48, 57), (10, 55, 77), (8, 51, 55), (11, 62, 67)}. The first data point, for example, represents a 10-year-old boy who is 59 inches tall and weighs 71 pounds.

The solution to this multiple linear regression problem is implemented by the program in *Listing 6.11*. This is similar to the quadratic regression program in *Listing 6.10*.

```java
public class Example5 {
    static double[][] x = { {10,59}, {9,57}, {12,61}, {10,52}, {9,48},
                {10,55}, {8,51}, {11,62} };
    static double[] y = {71, 68, 76, 56, 57, 77, 55, 67};
    static double[] b;     // coefficients to be determined

    public static void main(String[] args) {
        double[][] a = new double[3][3];
        double[] w = new double[3];
        deriveNormalEquations(a, w);
        printNormalEquations(a, w);
        solveNormalEquations(a, w);
        printResults();
    }

    public static void deriveNormalEquations(double[][] a, double[] w) {
        int n = y.length;
        for (int i = 0; i < n; i++) {
            double xi0 = x[i][0];
            double xi1 = x[i][1];
            double yi = y[i];
            a[0][0] = n;
            a[0][1] = a[1][0] += xi0;
            a[0][2] = a[2][0] += xi1;
            w[0] += yi;
            a[1][1] += xi0*xi0;
            a[2][2] += xi1*xi1;
            a[1][2] = a[2][1] += xi0*xi1;
            w[1] += xi0*yi;
            w[2] += xi1*yi;
        }
    }
```

```
run:
      8x0 +    79x1 +    445x2 =     527
     79x0 +   791x1 +   4427x2 =    5254
    445x0 + 4427x1 +  24929x2 =   29543
f(s, t) = -5.75 + 1.55s + 1.01t
f(10, 59) = 69.5
f(9, 57) = 65.9
f(11, 64) = 76.1
```

Listing 6.11: Multiple Linear Regression Example

In particular, its `printNormalEquations()` and `solveNormalEquations()` methods are the same.

The output shows that the normal equations for this dataset are:

$8a + 79b + 445c = 527$

$79a + 791b + 4427c = 5254$

$445a + 4427b + 24929c = 29543$

Their solution is:

$a = -5.747$

$b = 1.548$

$c = 1.013$

Thus, the multilinear regression function that solves this problem is:

$$f(x,y) = -5.75 + 1.55x + 1.01y$$

We can see that function agrees fairly well with the first two data points:

$$f(10,59) = 69.5 \approx z_1 = 71$$
$$f(9,57) = 65.9 \approx z_2 = 68$$

It predicts, for example, that an 11-year-old boy of height 64 inches would weigh 76.1 pounds:

$$f(11,64) = -5.75 + 1.55(11) + 1.01(64) = 76.1$$

The Apache Commons implementation

The Apache Commons Math library is extensive. We have already used a few of its implementations in this book. The `org.apache.commons.math3.stat.regression.SimpleRegression` class was used in *Listing 6.1*.

The program in *Listing 6.12* shows how the `OLSMultipleLinearRegression` class can be used to solve the previous problem. The complete program takes only 30 lines. The prefix **OLS** stands for **ordinary least squares**.

```java
import org.apache.commons.math3.stat.regression.OLSMultipleLinearRegression;

public class Example6 {
    static double[][] x = {{10, 59}, {9, 57}, {12, 61}, {10, 52}, {9, 48},
            {10, 55}, {8, 51}, {11, 62}};
    static double[] y = {71, 68, 76, 56, 57, 77, 55, 67};
    static double[] b;

    public static void main(String[] args) {
        OLSMultipleLinearRegression mlr = new OLSMultipleLinearRegression();
        mlr.newSampleData(y, x);
        b = mlr.estimateRegressionParameters();
        System.out.printf("f(s, t) = %.2f + %.2fs + %.2ft%n",
                b[0], b[1], b[2]);
        System.out.printf("f(10, 59) = %.1f%n", f(10, 59));
        System.out.printf("f(9, 57) = %.1f%n", f(9, 57));
        System.out.printf("f(11, 64) = %.1f%n", f(11, 64));
    }

    private static double f(double x1, double x2) {
        return b[0] + b[1]*x1 + b[2]*x2;
    }
}
```

```
run:
f(s, t) = -5.75 + 1.55s + 1.01t
f(10, 59) = 69.5
f(9, 57) = 65.9
f(11, 64) = 76.1
```

Listing 6.12: Using the OLSMultipleLinearRegression class

At line 18, the `newSampleData()` method loads the data into the `mlr` object. There are two versions of this method; the other one has three parameters.

At line 19, the `estimateRegressionParameters()` method returns the coefficient array `b[]`. All that remains is to print the results, which agree with those in *Listing 6.11*.

Curve fitting

Most relationships among variables are nonlinear. Among least squares solutions to such a relationship, there are three common approaches:

- Use polynomial regression
- Transform the given problem into an equivalent linear problem and then use linear regression
- Use least squares curve fitting

We have already demonstrated the first option. In this section, we will demonstrate the other two with examples.

As an example of transforming a nonlinear problem into a linear one, consider the standard equation for exponential decay:

$$y = y_0 e^{-rx}$$

Here, y_0 is the initial value of y and r is the rate of decay. (The constant e is the base of the natural logarithm, $e \approx 2.71828$.) Suppose we have a dataset relating the two variables x and y. The transformation to apply here is the natural logarithm. By taking the *ln* of both sides of the equation, we have:

$$\ln y = -rx$$

Substituting z for *ln y* gives us:

$$z = -rx$$

This is a linear equation. We can transform the dataset the same way, obtaining a dataset of the same size relating z to x. Then, we can apply simple linear regression to that data to obtain a solution of the form:

$$z = b_0 + b_1 x$$

Then, since $z = \ln y$, $y = e^z$, so we have:

$$y = e^z = e^{b_0 + b_1 x} = e^{b_0} e^{b_1 x}$$

The constant $e^{b_0} = y_0$, because that is the value of y where $x = 0$. So, our solution is:

$$y = e^{b_0} e^{b_1 x}$$

Here, b_0 and b_1 are the constants obtained from the linear regression algorithm.

It is important to note here that this solution is a least squares solution to the logarithmic formula. That is, the constants b_0 and b_1 minimize the objective function:

$$\sum (z_i - b_0 - b_1 x_i)^2$$

However, that is not the same as minimizing:

$$\sum (y_i - c_0 - e^{c_1 x})^2$$

The general method of least squares fitting can be used on more general, non-polynomial functions. For example, one might have a dataset that suggests the underlying function to be:

$$f(x) = b_0 + b_1 x + b_2^x$$

It could even be something like this:

$$f(x) = b_0 + b_1 x + b_2 x^2 + b_3^x + x b_4^x$$

You could apply the same calculus technique to minimize the objective function $\sum_{i=1}^{n} (y_i - f(x_i))^2$ by setting its partial derivatives equal to zero to derive the normal equations. But those equations would not be linear, so we could not solve them by methods such as LU decomposition.

There are numerical methods for solving systems of nonlinear equations. For bounded non-asymptotic functions, Fourier series or other orthogonal series are usually used, but those algorithms are beyond the scope of this book.

The Apache Commons Math library (`org.apache.commons.math3`) includes classes for curve fitting. You can choose a general curve type from the `math3.analysis.UnivariateFunction` interface and then apply it to a `math3.fitting.SimpleCurveFitter` object. For example, if your data is asymptotic at both ends, then the `math3.analysis.UnivariateFunction.Logistic` function could be a good choice. That might be best for medical data modelling the growth of tumors or economic data for a diffusion model.

Summary

In this chapter, we have investigated several examples of regression analysis, including linear regression, polynomial regression, multiple linear regression, and more general curve fitting. In each case, the objective is to derive from a given dataset a function that can then be used to extrapolate from the given data to predict unknown values of the function.

We've seen that these regression algorithms work by solving a system of linear equations, called the normal equations, for the problem. That part of the solution can be done by various algorithms, such as Cramer's Rule, Gaussian Elimination, or LU decomposition.

We used several approaches to implement these algorithms, including Windows Excel, direct Java implementations, and the Apache Commons Math library.

7
Classification Analysis

In the context of data analysis, the main idea of classification is the partition of a dataset into labeled subsets. If the dataset is a table in a database, then this partitioning could amount to no more than the addition of a new attribute (that is, a new table column) whose domain (that is, range of values) is a set of labels.

For example, we might have the table of 16 fruits shown in *Table 7-1*:

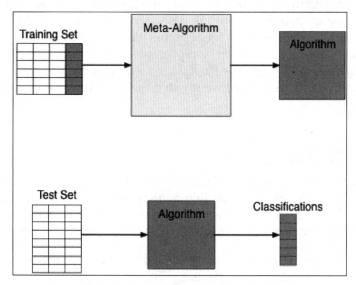

Figure 7-1. The meta-algorithm generates the algorithm

The last column, labeled Sweet, is a nominal attribute that can be used to classify fruit: either it's sweet or it isn't. Presumably, every fruit type that exists could be classified by this attribute. If you see an unknown fruit in the grocery store and wonder whether it is sweet, a classification algorithm could predict the answer, based upon the other attributes that you can observe {Size, Color, Surface}. We will see how to do that later in the chapter.

A classification algorithm is a **meta-algorithm**: its purpose is to generate a specific algorithm that can then be used to classify new objects. The input to the meta-algorithm is a training set, which is a sample dataset whose values are used to compute the parameters of the specific algorithm generated. We will use the dataset in *Table 7-1* as the input training set for several of the meta-algorithms in this chapter.

The relationship between a meta-algorithm and the algorithm that it generates is illustrated in *Figure 7-1*. The meta-algorithm takes the training set for input and produces the algorithm as output. That algorithm then takes a test dataset for input and produces a classification for each data point in the test set. Note that the data in the training set include values for the target attribute (for example, `Sweet`), but those in the test set do not.

It is possible to use the training set as a test set, usually to compare the predicted values (the classifications) with the corresponding vales in the training set. This is a standard method for measuring the accuracy of the algorithm. If we know the right answer, the algorithm should predict it.

The training set S is an ordinary dataset with $n + 1$ attributes $\{A_1, A_2, \ldots, A_n, T\}$, where T is the target attribute. In most cases, the A_j will be either all nominal or all numeric, depending upon the algorithm. Also, in most cases, T will be nominal, and usually Boolean. We think of S as a set of points with $n + 1$ coordinates: $(x_1, x_2, \ldots, x_n, y)$. The classification algorithm can then be represented as a function f, where the output $y = f(x_1, x_2, \ldots, x_n)$, or more simply, $y = f(x_1, x_2, \ldots, x_n) = f(x)$, where $x = (x_1, x_2, \ldots, x_n)$, is a typical element of the test set. Then we say that the algorithm predicts the value y for the data x.

Decision trees

A **decision tree** is a tree structure that is used like a flow-chart. Each internal node contains a true-false classification question, with its two branches corresponding to the answers *True* and *False*. Each leaf node is labeled with the consequent classification label. The paths from the root to the leaves represent classification rules.

The purpose of the decision tree is to provide a dynamic sequence of questions, whose answers will lead to a single decision. This is one type of classification algorithm. The target attribute is the set of all possible decisions that label the leaves of the tree.

The **Binary Search** algorithm is essentially a decision tree. It is used, for example, for looking up a name in a long, alphabetized list of names, such as a phone book or a contacts list. The process begins at the middle of the list and then branches repeatedly according to how the target name compares to the name in the current node of the tree.

This is illustrated in *Figure 7-2*:

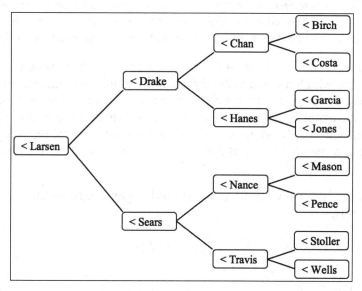

Figure 7-2. Binary search decision tree

Suppose you are looking for the name Kelly. Comparing it to **Larsen** at the root sends you up to the next node **Drake**, then down to **Hanes**, and down to **Jones**, continuing until you finally narrow it down to your target, Kelly.

What does entropy have to do with it?

To implement the standard classification algorithm for generating a decision tree, we must first understand the concept of the entropy of a partition. We approach that bridge by first reviewing elementary probability theory.

Recall from *Chapter 4, Statistics* the definitions of sample space, random variable, and probability function. For example, in the experiment of tossing a red die and a green die, the sample space S has 36 elementary outcomes, such as (2, 3) (meaning that the red die came up a two and the green die came up a three). If the dice are balanced, then each of those 36 outcomes is equally likely, and thus has the probability *p = 1/36*.

Probability events are sets of elementary outcomes. The probability of an event is the sum of the probabilities of its elementary outcomes. For example, the event that the sum of the two dice is 5 is the event E = {(1,4), (2,3), (3,2), (4,1)}. Its probability is $P(E)$ = 4/36 = 1/9.

Classification Analysis

The idea of entropy developed with the study of thermodynamics in the nineteenth century. In 1948, *Claude Shannon* applied the idea to information theory as a measure of the information content of a transmitted message. More recently, it has become important in data science. We explain it here in terms of elementary probability.

Given a sample space S, we wish to define a function `I( )` that assigns a number to each event $E \subseteq S$, which measures the information content of that event-in the sense of how much knowledge it conveys about what actually occurred in the experiment. For example, the event E in the dice experiment reports that the sum of the dice was 5, but it doesn't tell exactly what number came up on each of the two dice. So that $I(E)$ would be less than $I(E')$, if E' is the event (2, 3), because that event has more information.

This function `I()` that we want to define should depend only on the probabilities of the events; that is,

$$I(E) = f(P(E))$$

For some function f.

We can derive this f function from the properties that we know it must have. For example, $f(1) = 0$, because when $P(E) = 1$, the event is certain (100% probability), and that has 0 information content.

A more fruitful observation is this: the knowledge that two independent events happened should be the sum of the knowledge of each one:

$$I(E \cap F) = I(E) + I(F)$$

Think of paying someone for information about two events. If they are independent, then the amount you would pay for the information about both should be the sum of the amounts you'd pay for each.

Surprisingly, now, from that observation, we can derive the unknown function $f(\ )$ and thus obtain our definition of the entropy function $I(\ )$. This comes from the other fact we learned in *Chapter 4, Statistics*, that for independent events:

$$P(E \cap F) = P(E) \cdot P(F)$$

For example, $P(2,3) = P(2 \text{ on red}) \cdot P(3 \text{ on green}) = (1/6) \cdot (1/6) = 1/36$.

So, for any two independent events E and F:

$$f(P(E)) + f(P(F)) = I(E) + I(F)$$
$$= I(E \cap F)$$
$$= f(P(E \cap F))$$
$$= f(P(E) \cdot P(F))$$

Therefore, for any two probabilities, p and q:

$$f(p) + f(q) = f(p \cdot q)$$

In mathematics, it can be proved that the only functions that have that property are logarithmic functions:

$$\log(xy) = \log x + \log y$$

We'll specify the base of the logarithm later. The property holds for any base.

So now, we can conclude that the entropy function must be:

$$I(E) = K \log(P(E))$$

For some constant K. But I(E) should be a positive number, since it is a measure of information content, while $P(E) \leq 1$ because it is a probability, so $log\ (P(E)) \leq 0$. That means that the constant K must also be negative, say K = −K', for some positive constant K'. Then:

$$I(E) = -K' \log(P(E)) = -\lg(P(E))$$

So, the entropy of a single outcome x with probability p is:

$$I(x) = -\lg(p)$$

Here, lg stands for log_2, the binary (base 2) logarithm.

Classification Analysis

> Remember that all logarithms are proportional; for example, $\lg x = (\ln x)/(\ln 2) = 1.442695 \ln x$

This function $l()$ is a function on the sample space S itself (the set of all elementary outcomes), which makes it a random variable.

The **expected value** $E(X)$ of any random variable is the sum,

$$E(X) = \sum x p(x)$$

where the summation is over all values x of X. For example, if you play a game in which you toss one die and then get paid $\$x^2$ dollars when the number x comes up, then your expected payoff is:

$$E(X) = \$\sum x^2 p(x) = \frac{\$21}{6} = \$3.50$$

In other words, it would be a fair game if you paid $3.50 to play it.

Finally, we can define the **entropy** $H(S)$ of a sample space S to the expected value of the entropy random variable that we derived previously:

$$H(S) = \sum I(x) p(x) = \sum \left(-\lg(p(x))\right) p(x) = -\sum \left(\lg(p)\right) p$$

or, more simply,

$$H(S) = -\sum_i p_i \lg p_i$$

where $p_i = p(x_i)$. The sum is over all p_i that are greater than 0.

As an example, suppose you toss a biased coin, whose probability of heads is $p_0 = 3/4$. Then:

$$H(S) = -\sum_i p_i \lg p_i = -p_0 \lg p_0 - p_1 \lg p_1 = -\left(\frac{3}{4}\right)\lg\left(\frac{3}{4}\right) - \left(\frac{1}{4}\right)\lg\left(\frac{1}{4}\right) = 0.81$$

On the other hand, if the coin were fair, with both $p_i = \frac{1}{2}$, then:

$$H(S) = -\left(\frac{1}{2}\right)\lg\left(\frac{1}{2}\right) - \left(\frac{1}{2}\right)\lg\left(\frac{1}{2}\right) = -\left(\frac{1}{2}\right)(-1) - \left(\frac{1}{2}\right)(-1) = 1.00$$

In general, if S is a equiprobable space, where all $p_i = 1/n$, then:

$$H(S) = -\sum_i p_i \lg p_i = -\sum_i \frac{1}{n}\lg\frac{1}{n} = \sum_i \frac{\lg n}{n} = \frac{\lg n}{n}\sum_i 1 = \frac{\lg n}{n}(n) = \lg n$$

In the previous example, $n = 2$, so $H(S) = \lg 2 = 1$.

Entropy is a measure of uncertainty or randomness. The most uncertainty is when all outcomes are equally likely. Therefore, the maximum possible entropy for a space of n possible outcomes is $H(S) = \lg n$.

The other extreme would be a space with no uncertainty, where one $p_k = 1$ and all the other $p_i = 0$. In that case, $H(S) = -\sum_i p_i \lg p_i = -p_k \lg p_k = -(1)\lg 1 = -(1)(0) = 0$.

In general:

$$0 \leq H(S) \leq \lg n$$

The ID3 algorithm

Our main reason for developing the formula for entropy in the last section was to support the implementation of the ID3 algorithm for generating decision trees.

The ID3 algorithm generates an optimal decision tree from a dataset that has only nominal attributes. It was invented by J. R. Quinlan in 1986. The name stands for **Iterative Dichotomizer 3**.

The algorithm builds the decision tree from a training set, which is a relatively small subset of the entire population to which the tree will be applied. The correctness of the decision tree will depend upon how representative the training set is, much like a random sample from a population.

We will describe the ID3 algorithm by means of an example. Suppose we want a decision tree that will predict whether a previously unknown fruit is sweet or sour. The tree will predict the answer based upon the observed attributes of the fruit: Color, Size, and type of surface (SMOOTH, ROUGH, or FUZZY).

Classification Analysis

So, we begin with these specifications. The target attribute is:

```
Sweet = {T, F}
```

and the known attributes are:

```
Color = {RED, YELLOW, BLUE, GREEN, BROWN, ORANGE}
Surface = {SMOOTH, ROUGH, FUZZY}
Size = {SMALL, MEDIUM, LARGE}
```

The data in *Table 7-1* will be our training set. It has 16 data points, with 11 having T (true) for the Sweet attribute.

The algorithm will build the tree recursively. At each stage of the algorithm, it will determine the optimal attribute, among those remaining, for splitting the flow of the decision tree to the next level. That splitting strategy will be based upon the gain in entropy by that strategy, where gain from splitting on an attribute A is computed by the function:

$$Gain(A) = H(S) - \sum_i p_i H(a_i)$$

where $H(\cdot)$ is our entropy function, p_i is the proportion of training set data points whose value for attribute A is a_i, and the summation is over all values of the attribute A.

Name	Size	Color	Surface	Sweet
apple	MEDIUM	RED	SMOOTH	T
apricot	MEDIUM	YELLOW	FUZZY	T
banana	MEDIUM	YELLOW	SMOOTH	T
cherry	SMALL	RED	SMOOTH	T
coconut	LARGE	BROWN	FUZZY	T
cranberry	SMALL	RED	SMOOTH	F
fig	SMALL	BROWN	ROUGH	T
grapefruit	LARGE	YELLOW	ROUGH	F
kumquat	SMALL	ORANGE	SMOOTH	F
lemon	MEDIUM	YELLOW	ROUGH	F
orange	MEDIUM	ORANGE	ROUGH	T
peach	MEDIUM	ORANGE	FUZZY	T
pear	MEDIUM	YELLOW	SMOOTH	T
pineapple	LARGE	BROWN	ROUGH	T

Name	Size	Color	Surface	Sweet
pumpkin	LARGE	ORANGE	SMOOTH	F
strawberry	SMALL	RED	ROUGH	T

Table 7-1. Training Set for Fruit Example

[The entropy of a partition, $S = \{S_1, S_2, \ldots, S_n\}$ is defined the same as for a sample space, using the relative sizes for the probabilities: $p_i = |S_i|/|S|$. Also, if $S = \{S_1\}$ (that is, $n = 1$), then $E(S) = 0$.]

Suppose we begin by considering a split on the *Size* attribute. That would make the decision tree look like *Figure 7-3*. Notice that all the sweet fruits are underlined:

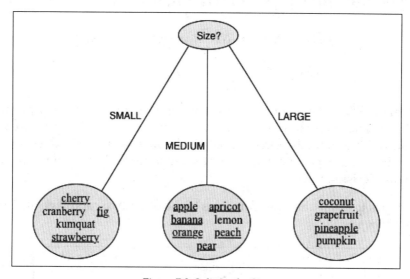

Figure 7-3: Splitting by Size

This would split *S* into three categories: *SMALL*, *MEDIUM*, and *LARGE*, partitioning our training data into nodes of size 5, 7, and 4. The proportions of sweet fruit in each are $p_1 = 3/5$, $p_2 = 6/7$, and $p_3 = 2/4$. The proportion of sweet fruits among the whole training set is 11/16. So, our calculations for *Gain(Size)* are:

$$H(S) = -\frac{11}{16}\lg\frac{11}{16} - \frac{5}{16}\lg\frac{5}{16} = 0.8960$$

Classification Analysis

For each of the three categories:

$$H(SMALL) = -\frac{3}{5}\lg\frac{3}{5} - \frac{2}{5}\lg\frac{2}{5} = 0.9710$$

$$H(MEDIUM) = -\frac{6}{7}\lg\frac{6}{7} - \frac{1}{7}\lg\frac{1}{7} = 0.591$$

$$H(LARGE) = -\frac{2}{4}\lg\frac{2}{4} - \frac{2}{4}\lg\frac{2}{4} = 1.0000$$

Thus, the gain for this split is:

$$\text{Gain}(\textbf{Size}) = H(S) - \left(\frac{5}{16}H(SMALL) + \frac{7}{16}H(MEDIUM) + \frac{4}{16}H(LARGE)\right)$$

$$= 0.8960 - \left(\frac{5}{16}(0.9710) + \frac{7}{16}(0.5917) + \frac{4}{16}(1.0000)\right)$$

$$= 0.0838$$

This gain, about 8%, is small, meaning that this split would not be very productive. We could have guessed that, since the 11 sweet fruits in our training set are nearly evenly distributed among the three categories. In other words, *Size* does not distinguish between sweet and sour very well, so it probably should not be the first splitter in our decision tree.

The next step is to apply the same formulas to the other two candidate attributes, *Color* and by *Surface*:

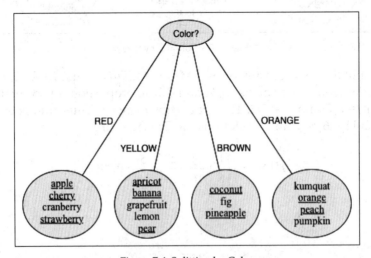

Figure 7-4: Splitting by Color

Chapter 7

For *Color*, the fractions of sweet fruit in each category are 3/4, 3/5, 2/3, and 2/4. So, the gain for splitting by *Color* is:

$$\begin{aligned}
\text{Gain}(\textbf{Color}) &= H(S) - \left(\frac{4}{16}H(\text{RED}) + \frac{5}{16}H(\text{YELLOW}) + \frac{4}{16}H(\text{ORANGE})\right) \\
&= 0.8960 - \left(\frac{4}{16}(0.8113) + \frac{5}{16}(0.9710) + \frac{3}{16}(0.9183) + \frac{4}{16}(1.0000)\right) \\
&= 0.0260
\end{aligned}$$

This gain, about 2%, is worse than the gain for splitting by `Size`.

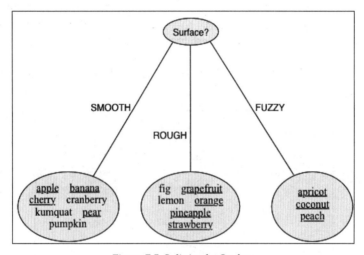

Figure 7-5: Splitting by Surface

For the *Surface* attribute, the fractions of sweet fruit in each category are 4/7, 4/6, and 3/3. This split looks more promising because, as we can see, it already determines that "*Fuzzy fruit is sweet.*":

$$\begin{aligned}
\text{Gain}(\textbf{Surface}) &= H(S) - \left(\frac{7}{16}H(\text{SMOOTH}) + \frac{6}{16}H(\text{ROUGH}) + \frac{3}{16}H(\text{FUZZY})\right) \\
&= 0.8960 - \left(\frac{7}{16}(1.9852) + \frac{6}{16}(0.9183) + \frac{3}{16}(0) + \frac{4}{16}(1.0000)\right) = 0.1206
\end{aligned}$$

This is about a 12% gain—much better than 8% or 2%. So, our first decision is to split by the `Surface` attribute at the root node.

Classification Analysis

Notice that in the preceding calculations, the highest entropy value was 1.0000, for *H(LARGE)* in the *Gain(Size)* calculation and for *H(ORANGE)* in the *Gain(Color)* calculation. That was because, in both cases, exactly half (2 out of 4) of the fruits in the subcategory are sweet. That value, *1.0*, is the maximum possible entropy value for a binary attribute (in this case, *Sweet*). It reflects the maximal uncertainty, or lack of information, of that node. Guessing whether a fruit is sweet in that subcategory (*ORANGE* in *Color* or *LARGE* in *Size*) might as well be done at random.

On the other hand, in the *Gain(Surface)* calculation, *H(FUZZY) = 0*, is the lowest possible entropy value. That reflects the fact that there is no uncertainty in this category: all the fruit are sweet. It's the best kind of subcategory to have in a decision tree.

Now that the algorithm has chosen to split first on the `Surface` attribute, the next step is to examine the gains from each of the other two attributes at each of the three second level nodes, for the subcategories SMOOTH, ROUGH, and FUZZY:

First, consider the two possibilities for the SMOOTH branch emanating from the root in *Figure 7.5*:

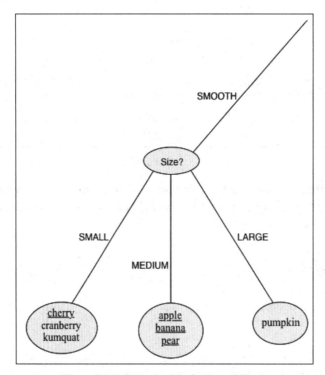

Figure 7-6: Splitting by Size for Smooth Fruit

We can split by Size or by Color.

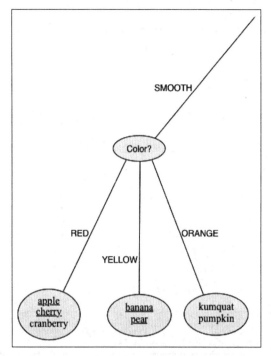

Figure 7-7: Splitting by Color for Smooth Fruit

The relevant proportions for splitting by *Size* are: 4 out of the 7 fruits are sweet, 1 out of the 3 *SMALL* fruits are sweet, 3 out of the 3 *MEDIUM* sized fruits are sweet, and 0 out of the 1 the *LARGE* fruit is sweet, as *Figure 7.6* shows. So, the gain for splitting the *SMOOTH* fruit by *Size* is:

$$\text{Gain}(\textbf{Size} \mid \text{SMOOTH}) = 0.9852 - \left(\frac{3}{7}(0.9183) + \frac{3}{7}(0) + \frac{1}{7}(0)\right) = 0.5917$$

The relevant proportions for splitting by Color are: 4/7, 2/3, 2/2, and 0/2, as *Figure 7-7* shows. So, the gain for splitting the SMOOTH fruit by Color is:

$$\text{Gain}(\textbf{Color} \mid \text{SMOOTH}) = 0.9852 - \left(\frac{3}{7}(0.9183) + \frac{2}{7}(0) + \frac{2}{7}(0)\right) = 0.5917$$

It's a tie! Both *Size* and *Color* are equally good discriminators for *SMOOTH* fruit (according to our training set). You can see why. In both cases, only the first subcategory is ambiguous. For both *Size* and *Color*, the first subcategory is split 2 to 1, the second subcategory is all sweet fruit, and the third is all sour.

Classification Analysis

At this point, the ID3 algorithm would flip a coin. We'll assume that it chose the Size attribute for splitting the *SMOOTH* fruit node.

Next, the algorithm considers the two possibilities for the *ROUGH* branch from *Figure 7-5*:

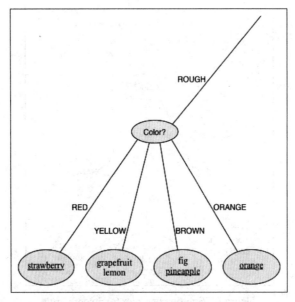

Figure 7-8: Splitting by Color for Rough Fruit

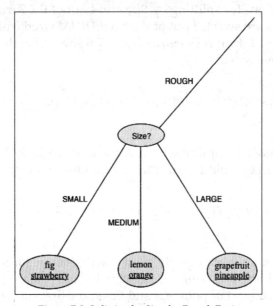

Figure 7-9: Splitting by Size for Rough Fruit

The relevant proportions for splitting by *Size* are: 4/6, 2/2, 1/2, and 1/2, as *Figure 7-8* shows. So, the gain for splitting the *ROUGH* fruit by *Size* is:

$$\text{Gain}(\textbf{Size}\,|\,\text{ROUGH}) = 0.9183 - \left(\frac{2}{6}(0) + \frac{2}{6}(1.0000) + \frac{2}{6}(1.0000)\right) = 0.2516$$

The relevant proportions for splitting by *Color* are: 4/6, 1/1, 0/2, 2/2, and 1/1, as *Figure 7-9* shows. So, the gain for splitting the *ROUGH* fruit by *Color* is:

$$\text{Gain}(\textbf{Color}\,|\,\text{ROUGH}) = 0.9183 - \left(\frac{1}{6}(0) + \frac{2}{6}(0) + \frac{2}{6}(0) + \frac{1}{6}(0)\right) = 0.9183$$

Obviously, *Color* wins this contest. That should be evident from *Figure 7-9*. That sub-tree provides a complete, unambiguous categorization of fruit with a *ROUGH* surface: if it's *YELLOW*, it's sour, otherwise, it's sweet (again, according to our training set).

The only remaining sub-categorization now is for *SMOOTH, SMALL* fruit (see *Figure 7-6*). The only remaining attribute at that node is *Color*. Of the three fruits {cherry, cranberry, kumquat} in that node, cherry and cranberry are RED, and kumquat is ORANGE. So, the algorithm will assign the value T for the *Sweet* attribute for RED fruits and it will assign the value F for ORANGE fruits.

The final form of the complete decision tree is shown in *Figure 7-10*.

This result is not perfect in its categorization of the 16 fruits in our training set. It gives the wrong answer for cranberry. But that is because the training set itself is insufficient. It lists two fruits that are SMALL, RED, and SMOOTH: cranberry and cherry. But one is sweet and the other isn't!

The algorithm has two tie-breaking strategies. One kind of tie is when two subcategories have the same Gain() value, as we had with Size and Color of SMOOTH fruit. That kind of tie is resolved with a random choice.

Classification Analysis

The other kind of tie is when the algorithm reaches a leaf of the tree, with no more attributes for further splitting, and yet the node still has an equal number of elements of each value of the target attribute. That's what happened at the SMOOTH SMALL RED node, the one containing {cherry, cranberry}. The resolution to this kind of tie is to assign the value owned by a majority of all the data points in the training set. In this set, 11 of the 16 points have the value T for Sweet.

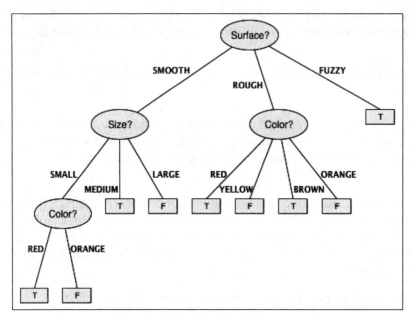

Figure 7-10: The complete decision tree

The order in which the recursive ID3 algorithm will create the nodes of the decision tree is not the same as we indicated here in this Fruit example. It will, instead, adhere to the preorder traversal of the tree. That order is indicated in *Figure 7-11*:

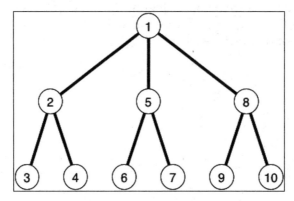

Figure 7.11: Preorder traversal of tree

This order begins at the root **node (1)**, then to its left-most child **node (2)**, and then to its left-most child **node (3)**. This process continues until a leaf node is reached. The order then continues with that nodes sibling **node (4)**, and on until all of them have been created. Then the order continues with the next sibling **node (5)** of the parent, and so on.

At each node in a preorder traversal, every ancestor node and every node on its left will have already been processed.

Java Implementation of the ID3 algorithm

A complete Java implementation of the ID3 algorithm is beyond the scope of this book. However, we can see how Java helped implement the algorithm on our Fruit training set.

Listing 7.1 shows the three utility methods that we used for the calculations. The binary logarithm is defined at lines 78-80. It simply implements the formula:

$$\log_2 x = \frac{\ln x}{\ln 2}$$

```java
/*  Gain for the splitting {A1, A2, ...}, where Ai
    has n[i] points, m[i] of which are favorable.
 */
public static double g(int[] m, int[] n) {
    int sm = 0, sn = 0;
    double nsh = 0.0;
    for (int i = 0; i < m.length; i++) {
        sm += m[i];
        sn += n[i];
        nsh += n[i]*h(m[i],n[i]);
    }
    return h(sm, sn) - nsh/sn;
}

/*  Entropy for m favorable items out of n.
 */
public static double h(int m, int n) {
    if (m == 0 || m == n) {
        return 0;
    }
    double p = (double)m/n, q = 1 - p;
    return -p*lg(p) - q*lg(q);
}

/*  Returns the binary logarithm of x.
 */
public static double lg(double x) {
    return Math.log(x)/Math.log(2);
}
```

Listing 7.1: Utility methods for the ID3 algorithm

Classification Analysis

The entropy function is defined as `h(m,n)` at lines 68-74. It implements the formula:

$$h = -\left(\frac{m}{n}\right)\lg\left(\frac{m}{n}\right) - \left(\frac{n-m}{n}\right)\lg\left(\frac{n-m}{n}\right)$$

The `Gain()` function is defined as `g(m[],n[])` at lines 55-64. It implements the formula:

$$\text{Gain}(A) = H(S) - \sum_i p_i H(a_i)$$

The code in *Listing 7.2* shows how these methods were used for the first few computations for the preceding Fruit example:

```
Trainer.dat   ComputeGain.java
10  public class ComputeGain {
11      public static void main(String[] args) {
12          System.out.printf("h(11,16) = %.4f%n", h(11,16));
13          System.out.println("Gain(Size):");
14          System.out.printf("\th(3,5) = %.4f%n", h(3,5));
15          System.out.printf("\th(6,7) = %.4f%n", h(6,7));
16          System.out.printf("\th(2,4) = %.4f%n", h(2,4));
17          System.out.printf("\tg({3,6,2},{5,7,4}) = %.4f%n",
18              g(new int[]{3,6,2},new int[]{5,7,4}));
19          System.out.println("Gain(Color):");
20          System.out.printf("\th(3,4) = %.4f%n", h(3,4));
21          System.out.printf("\th(3,5) = %.4f%n", h(3,5));
22          System.out.printf("\th(2,3) = %.4f%n", h(2,3));
23          System.out.printf("\th(2,4) = %.4f%n", h(2,4));
24          System.out.printf("\tg({3,3,2,2},{4,5,3,4}) = %.4f%n",
25              g(new int[]{3,3,2,2},new int[]{4,5,3,4}));

Output - DecisionTrees (run)
run:
h(11,16) = 0.8960
Gain(Size):
        h(3,5) = 0.9710
        h(6,7) = 0.5917
        h(2,4) = 1.0000
        g({3,6,2},{5,7,4}) = 0.0838
Gain(Color):
        h(3,4) = 0.8113
        h(3,5) = 0.9710
        h(2,3) = 0.9183
        h(2,4) = 1.0000
        g({3,3,2,2},{4,5,3,4}) = 0.0260
```

Listing 7.2: First few computations for fruit example

The first computation, at line 12, is for the entropy of the entire training set, S. It is *h(11,16)* because 11 of the 16 data points had the favorable value for the target attribute (T for Sweet):

$$H(S) = -\frac{11}{16}\lg\frac{11}{16} - \frac{5}{16}\lg\frac{5}{16} = 0.8960$$

At line 14, we compute the entropy for the *SMALL* subcategory in the *Size* partition:

$$H(\text{SMALL}) = -\frac{3}{5}\lg\frac{3}{5} - \frac{2}{5}\lg\frac{2}{5} = 0.9710$$

This is h(3,5) because 3 of the 5 SMALL fruits are sweet.

At line 18, we compute:

$$\text{Gain}(\textbf{Size}) = H(S) - \left(\frac{5}{16}H(\text{SMALL}) + \frac{7}{16}H(\text{MEDIUM}) + \frac{4}{16}H(\text{LARGE})\right)$$

$$= 0.8960 - \left(\frac{5}{16}(0.9710) + \frac{7}{16}(0.5917) + \frac{4}{16}(1.0000)\right)$$

$$= 0.0838$$

This is computed as:

```
g(new int[]{3,6,2}, new int[]{5,7,4})
```

The first array {3,6,2} is m[], and the second array {5,7,4} is n[]. These values were used because the proportions of sweet fruit among the SMALL, MEDIUM, and LARGE categories are 3/5, 6/7, and 2/4, respectively (see *Figure 7-3*).

Similarly, the expression

```
g(new int[]{3,3,2,2}, new int[]{4,5,3,4})
```

is used at line 25 to compute Gain(Color) = 0.0260. The arrays {3,3,2,2} and {4,5,3,4} were used because the proportions of sweet fruit among the RED, YELLOW, BROWN, and ORANGE categories are 3/4, 3/5, 2/3, and 2/4, respectively (see *Figure 7.4*).

Classification Analysis

The Weka platform

Weka is a collection of Java libraries and tools that implement machine learning algorithms. It was developed and is maintained by computer scientists at the **University of Waikato** in New Zealand. Among these packages are implementations of most of the classification algorithms that we are considering in this chapter. Their download site is `http://www.cs.waikato.ac.nz/ml/weka/downloading.html`.

If you are using NetBeans on a Mac, copy the `weka.jar` file and the `data/` and `doc/` folders into `/Library/Java/Extensions/`. Then in NetBeans, select **Tools | Libraries** to bring up the **Ant Library Manager** panel. Under **Classpath** (the default), select **Add JAR/Folder** and navigate to your `weka` folder (for example, `/Library/Java/Extensions/weka-3-9-1/`), click on the **Add JAR/Folder** button, and then click **OK**.

The Weka Workbench, an online PDF about how to use Weka, can be read at `http://www.cs.waikato.ac.nz/ml/weka/Witten_et_al_2016_appendix.pdf`.

The ARFF filetype for data

Weka uses a specialized data format for its input files, called **Attribute-Relation File Format** (**ARFF**). You can read about it at `http://weka.wikispaces.com/ARFF`.

The file shown in *Listing 7-3* is a typical ARFF file. It is one of the examples provided in the WEKA data folder that you downloaded previously (we replaced 46 lines of comments with one at line 3).

As the website described, an ARFF file has three kinds of lines: comments, metadata, and data. Comment lines begin with the percent sign, `%`. Metadata lines begin with the "at" symbol, `(@)`. Data lines contain only CSV data, one data point per line.

You can think of an ARFF file as a structured CSV file, where the structure is defined by the metadata statements. Each metadata line begins with a keyword: `relation`, `attribute`, or `data`. Each attribute line specifies the name of an attribute, followed by its datatype:

```
% 1. Title: Database for fitting contact lenses
%
% <MORE COMMENTS>
%
% 9. Class Distribution:
%      1. hard contact lenses: 4
%      2. soft contact lenses: 5
%      3. no contact lenses: 15

@relation contact-lenses

@attribute age                  {young, pre-presbyopic, presbyopic}
@attribute spectacle-prescrip   {myope, hypermetrope}
@attribute astigmatism          {no, yes}
@attribute tear-prod-rate       {reduced, normal}
@attribute contact-lenses       {soft, hard, none}

@data
%
% 24 instances
%
young,myope,no,reduced,none
young,myope,no,normal,soft
young,myope,yes,reduced,none
young,myope,yes,normal,hard
young,hypermetrope,no,reduced,none
young,hypermetrope,no,normal,soft
young,hypermetrope,yes,reduced,none
young,hypermetrope,yes,normal,hard
pre-presbyopic,myope,no,reduced,none
pre-presbyopic,myope,no,normal,soft
pre-presbyopic,myope,yes,reduced,none
pre-presbyopic,myope,yes,normal,hard
pre-presbyopic,hypermetrope,no,reduced,none
pre-presbyopic,hypermetrope,no,normal,soft
pre-presbyopic,hypermetrope,yes,reduced,none
pre-presbyopic,hypermetrope,yes,normal,none
presbyopic,myope,no,reduced,none
presbyopic,myope,no,normal,none
presbyopic,myope,yes,reduced,none
presbyopic,myope,yes,normal,hard
presbyopic,hypermetrope,no,reduced,none
presbyopic,hypermetrope,no,normal,soft
presbyopic,hypermetrope,yes,reduced,none
presbyopic,hypermetrope,yes,normal,none
```

Listing 7-3. *An ARFF data file*

Classification Analysis

In the `contact-lenses.arff` file, all datatypes are nominal: their values are listed as a set of strings. The other possible datatypes are numeric, string, and date.

The Fruit data that we used in the ID3 example is shown in ARFF format in *Listing 7.4*. Notice that ARFF format is case insensitive with metadata, and it also ignores extra whitespace:

```
1  @RELATION Fruit
2  @ATTRIBUTE Name String
3  @ATTRIBUTE Size {SMALL,MEDIUM,LARGE}
4  @ATTRIBUTE Color {RED,YELLOW,BROWN,ORANGE,GREEN}
5  @ATTRIBUTE Surface {SMOOTH,ROUGH,FUZZY}
6  @ATTRIBUTE Sweet {F,T}
7  @DATA
8  apple        MEDIUM    RED       SMOOTH    T
9  apricot      MEDIUM    YELLOW    FUZZY     T
10 banana       MEDIUM    YELLOW    SMOOTH    T
11 cherry       SMALL     RED       SMOOTH    T
12 coconut      LARGE     BROWN     FUZZY     T
13 cranberry    SMALL     RED       SMOOTH    F
14 fig          SMALL     BROWN     ROUGH     T
15 grapefruit   LARGE     YELLOW    ROUGH     F
16 kumquat      SMALL     ORANGE    SMOOTH    F
17 lemon        MEDIUM    YELLOW    ROUGH     F
18 orange       MEDIUM    ORANGE    ROUGH     T
19 peach        MEDIUM    ORANGE    FUZZY     T
20 pear         MEDIUM    YELLOW    SMOOTH    T
21 pineapple    LARGE     BROWN     ROUGH     T
22 pumpkin      LARGE     ORANGE    SMOOTH    F
23 strawberry   SMALL     RED       ROUGH     T
```

Listing 7-4: Our Fruit Data in ARFF format

The little program in *Listing 7-5* shows how Weka manages data. The `Instances` class is like the Java `ArrayList` class or `HashSet` class, storing a collection of `Instance` objects, each representing a data point in the dataset. So, an `Instances` object is like a database table, and an `Instance` object is like a row in a database table.

This data file is opened in the program shown in *Listing 7-5* at line 14 as a `DataSource` object. Then at line 16, the complete dataset is read into the `instances` object. Line 17 identifies the target attribute as `Sweet` (at index 3). The third attribute, `Color` (at index 2) is printed at line 18. Then the fourth instance (at index 3) is named `instance` at line 20 and printed at line 21. This is the same data that was read from line 11 of the ARFF file in *Listing 7.4*. Finally, the values of its first and third fields (`cherry` and `RED`) are printed at lines 22-23:

```java
import weka.core.Instance;
import weka.core.Instances;
import weka.core.converters.ConverterUtils.DataSource;

public class TestDataSource {
    public static void main(String[] args) throws Exception {
        DataSource source = new DataSource("data/fruit.arff");

        Instances instances = source.getDataSet();
        instances.setClassIndex(instances.numAttributes() - 1);
        System.out.println(instances.attribute(2));

        Instance instance = instances.get(3);
        System.out.println(instance);
        System.out.println(instance.stringValue(0));
        System.out.println(instance.stringValue(2));
    }
}
```

```
run:
@attribute Color {RED,YELLOW,BROWN,ORANGE,GREEN}
cherry,SMALL,RED,SMOOTH,T
cherry
RED
```

Listing 7.5: Testing the instances and instance classes

Java implementation with Weka

Weka can construct the same decision tree that we built for the fruit training data. The only change we need is to remove the Name attribute from the data file because the classifier works only with nominal datatypes. We named the resulting reduced file `AnonFruit.arff`.

Classification Analysis

The program and its output are shown in *Listing 7-6*. The J48 class instantiated at line 19 extends the ID3 algorithm that we described previously. It builds the decision tree at line 21. Then the loop at lines 23-27 iterate over the same 16 fruits, printing (in numerical form) both the actual value of the Sweet attribute and the value that the decision tree predicts. You can see that the two values agree for every fruit except the fourth one, which you can see in *Listing 7.6* is the cherry data point. Its value is 1 in the data file, but the tree predicts 0.

```java
import weka.classifiers.trees.J48;
import weka.core.Instances;
import weka.core.Instance;
import weka.core.converters.ConverterUtils.DataSource;

public class TestWekaJ48 {
    public static void main(String[] args) throws Exception {
        DataSource source = new DataSource("data/AnonFruit.arff");
        Instances instances = source.getDataSet();
        instances.setClassIndex(3);  // target attribute: (Sweet)

        J48 j48 = new J48();  // an extension of ID3
        j48.setOptions(new String[]{"-U"});  // use unpruned tree
        j48.buildClassifier(instances);

        for (Instance instance : instances) {
            double prediction = j48.classifyInstance(instance);
            System.out.printf("%4.0f%4.0f%n",
                    instance.classValue(), prediction);
        }
    }
}
```

```
run:
   1   1
   1   1
   1   1
   1   0
   1   1
   0   0
   1   1
   0   0
   0   0
   0   0
   1   1
   1   1
   1   1
   1   1
   0   0
   1   1
```

Listing 7.6: The Weka J48 classifier for the Fruit data

Recall that when we constructed the decision tree in *Figure 7-10*, we ended up with one node containing both `cherry` and `cranberry`. That was because both fruits are SMOOTH, SMALL, and RED. We broke the tie by assigning T (sweet) to both because that was the majority value in the training set. Evidently, Weka's J48 class uses F (0) for the default value instead (`cranberry` is the sixth fruit). As we mentioned previously, this conflict would not occur if our training set were internally consistent.

Bayesian classifiers

The **naive Bayes classification** algorithm is a classification process that is based upon Bayes' Theorem, which we examined in *Chapter 4, Statistics*. It is embodied in the formula:

$$P(E|F) = \frac{P(F|E)P(E)}{P(F)}$$

where E and F are events with probabilities $P(E)$ and $P(F)$, is the conditional probability of E given that F is true, and $P(F|E)$ is the conditional probability of F given that E is true. The purpose of this formula is to compute one conditional probability, $P(E|F)$, in terms of its reverse conditional probability $P(F|E)$.

In the context of classification analysis, we assume the population of data points is partitioned into m disjoint categories, $C_1, C_2,..., C_m$. Then, for any data point **x** and any specified category C_i:

$$P(C_i|\mathbf{x}) = \frac{P(\mathbf{x}|C_i)P(C_i)}{P(\mathbf{x})}$$

The Bayesian algorithm predicts which category C_i the point **x** is most likely to be in; that is, finding which C_i maximizes $P(C_i|\mathbf{x})$. But we can see from the formula that that will be the same C_i that maximizes $P(\mathbf{x}|C_i)P(C_i)$, since the denominator $P(\mathbf{x})$ is constant.

So that's the algorithm: compute $P(\mathbf{x}|C_i)P(C_i)$, for each $i = 1, 2,..., n$, and then pick the largest among those n numbers. The point **x** is then classified as being in that category.

To put this algorithm in the context of our previous classification analysis, recall that we have a training set S, which is a relational table whose attributes $A_1, A_2,..., A_n$ are nominal. We can identify each attribute A_j with its set of possible values $\{a_{j1}, a_{j2},..., a_{jk}\}$. For example, the Size attribute that we used before identifies with its set of values {SMALL, MEDIUM, LARGE}.

Classification Analysis

The target attribute is usually the last one specified, A_n. In our previous example, that attribute was Sweet, with the attribute set {F, T}. The purpose of a classification algorithm is to predict, for any new data point x, what the value of the target attribute is; for example, is that fruit sweet?

In this context, the categories correspond to the values of the target attribute. So, for our Fruit example, there are two categories: C_1 = *set of all sweet fruit*, and C_2 = *set of all sour fruit*.

Our goal is to compute each $P(\mathbf{x}|C_i)P(C_i)$, for each $i = 1, 2, \ldots, n$, and then pick the largest. The probabilities are all computed as relative frequencies. So, each $P(C_i)$ will simply be the number of training set data points that are in category C_i divided by the total number of data points.

To compute each $P(\mathbf{x}|C_i)$, we remember that x is a vector, $\mathbf{x} = (x_1, x_2, \ldots, x_n)$, where each x_j is a value of the corresponding attribute A_j. For example, if x represents the fruit cola, then $\mathbf{x}$ = (SMALL, RED, SMOOTH) according to our training set. We assume that the attributes are all mutually independent, so that:

$$P(x_j \wedge x_k) = P(x_j)P(x_k)$$

For example, the probability of being SMALL and RED equals the probability of being SMALL times the probability of being RED. Consequently, we can compute each $P(\mathbf{x}|C_i)$ as

$$P(\mathbf{x}|C_i) = P(x_1|C_i)P(x_2|C_i)\cdots P(x_n|C_i)$$

where each $P(x_j|C_i)$ is simply the number of points in category C_i that have the value x_j for attribute A_j. For example, in our training set in *Figure 7-2*, if $\mathbf{x}$ = *apple*, then:

$$P(\mathbf{x}|C_i) = P(\text{SMALL}|C_i)P(\text{RED}|C_i)\cdots P(\text{SMOOTH}|C_i)$$

The conditional probability $P(SMALL\,|\,C_i)$ is the proportion of fruit in category C_i that are small. If C_i is the category of sweet fruit, then $P(SMALL\,|\,C_i)$ is the proportion of sweet fruit that are small, which is 3/11, according to our training set. In that case, the complete calculation is:

$$P(\mathbf{x}\,|\,C_i) = \left(\frac{3}{11}\right)\left(\frac{3}{11}\right)\left(\frac{4}{11}\right) = \frac{36}{1331}$$

To finish the algorithm, we must compute and compare $P(\mathbf{x}\,|\,C_1)P(C_1)$ and $P(\mathbf{x}\,|\,C_2)P(C_2)$, where C_1 and C_2 are the categories of sweet and sour fruit, respectively, and $\mathbf{x}$ = (SMALL, RED, SMOOTH):

$$P(\mathbf{x}\,|\,C_1)P(C_1) = \left(\frac{36}{1331}\right)\left(\frac{11}{16}\right) = 0.0186$$

$$P(\mathbf{x}\,|\,C_2) = \left(\frac{2}{5}\right)\left(\frac{1}{5}\right)\left(\frac{3}{5}\right) = \frac{6}{125}$$

$$P(\mathbf{x}\,|\,C_2)P(C_2) = \left(\frac{6}{125}\right)\left(\frac{5}{16}\right) = 0.0150$$

Suppose we want to predict whether the fruit *cola* is sweet. Since $P(\mathbf{x}\,|\,C_1)P(C_1) > P(\mathbf{x}\,|\,C_2)P(C_2)$, we know that $P(C_1\,|\,\mathbf{x}) > P(C_2\,|\,\mathbf{x})$, and therefore classify $\mathbf{x}$ as category C_1, predicting that cola fruit is sweet. In fact, fresh cola fruit is not sweet.

Classification Analysis

Java implementation with Weka

To implement the Bayes Classification algorithm, we tailor it to our specific Fruit example, using the data from our *Fruit.arff* file shown in *Listing 7-4*. The code in *Listing 7-7* defines a class that encapsulates the data instances from that source:

```java
public class Fruit {
    String name, size, color, surface;
    boolean sweet;

    public Fruit(String name, String size, String color, String surface,
            boolean sweet) {
        this.name = name;
        this.size = size;
        this.color = color;
        this.surface = surface;
        this.sweet = sweet;
    }

    @Override
    public String toString() {
        return String.format("%-12s%-8s-8s-8s%s",
                name, size, color, surface, (sweet? "T": "F") );
    }

    public static Set<Fruit> loadData(File file) {
        Set<Fruit> fruits = new HashSet();
        try {
            Scanner input = new Scanner(file);
            for (int i = 0; i < 7; i++) {  // read past metadata
                input.nextLine();
            }
            while (input.hasNextLine()) {
                String line = input.nextLine();
                Scanner lineScanner = new Scanner(line);
                String name = lineScanner.next();
                String size = lineScanner.next();
                String color = lineScanner.next();
                String surface = lineScanner.next();
                boolean sweet = (lineScanner.next().equals("T"));
                Fruit fruit = new Fruit(name, size, color, surface, sweet);
                fruits.add(fruit);
            }
        } catch (FileNotFoundException e) {
            System.err.println(e);
        }
        return fruits;
    }

    public static void print(Set<Fruit> fruits) {
        int k=1;
        for (Fruit fruit : fruits) {
            System.out.printf("%2d. %s%n", k++, fruit);
        }
    }
}
```

Listing 7-7. A Fruit class

The algorithm is then implemented in the program shown in *Listing 7-8*:

```java
public class BayesianTest {
    private static Set<Fruit> fruits;

    public static void main(String[] args) {
        fruits = Fruit.loadData(new File("data/Fruit.arff"));
        Fruit fruit = new Fruit("cola", "SMALL", "RED", "SMOOTH", false);
        double n = fruits.size();   // total number of fruits in training set
        double sum1 = 0;            // number of sweet fruits
        for (Fruit f : fruits) {
            sum1 += (f.sweet? 1: 0);
        }
        double sum2 = n - sum1;     // number of sour fruits
        double[][] p = new double[4][3];
        for (Fruit f : fruits) {
            if (f.sweet) {
                p[1][1] += (f.size.equals(fruit.size)? 1: 0)/sum1;
                p[2][1] += (f.color.equals(fruit.color)? 1: 0)/sum1;
                p[3][1] += (f.surface.equals(fruit.surface)? 1: 0)/sum1;
            } else {
                p[1][2] += (f.size.equals(fruit.size)? 1: 0)/sum2;
                p[2][2] += (f.color.equals(fruit.color)? 1: 0)/sum2;
                p[3][2] += (f.surface.equals(fruit.surface)? 1: 0)/sum2;
            }
        }
        double pc1 = p[1][1]*p[2][1]*p[3][1]*sum1/n;
        double pc2 = p[1][2]*p[2][2]*p[3][2]*sum2/n;
        System.out.printf("pc1 = %.4f, pc2 = %.4f%n", pc1, pc2);
        System.out.printf("Predict %s is %s.%n",
                fruit.name, (pc1 > pc2? "sweet": "sour"));
    }
}
```

```
run:
pc1 = 0.0186, pc2 = 0.0150
Predict cola is sweet.
```

Listing 7-8. Naive Bayes classifier

Weka provides a good implementation of the Weka naive Bayes algorithm. Most of the code is in their `weka.classifiers.bayes.NaiveBayes` package. Documentation is available at http://weka.sourceforge.net/doc.dev/weka/classifiers/bayes/NaiveBayes.html.

Classification Analysis

The program in *Listing 7-9* tests this classifier:

```java
import java.util.List;
import weka.classifiers.Evaluation;
import weka.classifiers.bayes.NaiveBayes;
import weka.classifiers.evaluation.Prediction;
import weka.core.Instance;
import weka.core.Instances;
import weka.core.converters.ConverterUtils.DataSource;

public class TestWekaBayes {
    public static void main(String[] args) throws Exception {
        DataSource source = new DataSource("data/AnonFruit.arff");
        Instances train = source.getDataSet();
        train.setClassIndex(3);   // target attribute: (Sweet)

        NaiveBayes model=new NaiveBayes();
        model.buildClassifier(train);

        Instances test = train;
        Evaluation eval = new Evaluation(test);
        eval.evaluateModel(model,test);
        List <Prediction> predictions = eval.predictions();

        int k = 0;
        for (Instance instance : test) {
            double actual = instance.classValue();
            double prediction = eval.evaluateModelOnce(model, instance);
            System.out.printf("%2d.%4.0f%4.0f", ++k, actual, prediction);
            System.out.println(prediction != actual? " *": "");
        }
    }
}
```

```
run:
 1.   1   1
 2.   1   1
 3.   1   1
 4.   1   1
 5.   1   1
 6.   0   1 *
 7.   1   1
 8.   0   0
 9.   0   0
10.   0   1 *
11.   1   1
12.   1   1
13.   1   1
14.   1   1
15.   0   0
16.   1   1
```

Listing 7-9. Weka's Naive Bayes classifier

The output shows the actual value and the predicted value (1 = sweet, 0 = sour) on each of the 16 fruit instances in our fruit file (*Listing 7-4*) with the names removed. Compare the output here with that in *Listing 7-6*. The Bayes classifier makes two wrong predictions; the J48 classifier got only one wrong.

Support vector machine algorithms

A more recent approach to machine classification is to use a **support vector machine** (**SVM**). This idea originated with *Vladimir Vapnik* in 1995. It applies to the classification of datasets into two categories; that is, the target attribute has only two values, such as F and T for our Sweet attribute.

The main idea is to view each data point in the training set as an element in *n-dimensional Euclidean* space, where *n* is the number of non-target attributes. Then the SVM algorithm seeks an optimal hyperplane that separates the set of points into the two categories.

For example, our (anonymous) Fruit training set has three non-target attributes: Size, Color, and Surface. So, to run the SVM algorithm on it requires mapping each of the 16 data points into $\mathbb{R}^3$ (three-dimensional Euclidean space). For example, the first data point: (apple, MEDIUM, RED, SMOOTH, T) will be represented by the point (2, 1, 1).

Table 7-2 shows the results of that mapping:

Name	Size	Color	Surface	Sweet
apple	2	1	1	T
apricot	2	2	3	T
banana	2	2	1	T
cherry	1	1	1	T
coconut	3	3	3	T
cranberry	1	1	1	F
fig	1	3	2	T
grapefruit	3	2	2	F
kumquat	1	4	1	F
lemon	2	2	2	F
orange	2	4	2	T
peach	2	4	3	T
pear	2	2	1	T

Classification Analysis

Name	Size	Color	Surface	Sweet
pineapple	3	3	2	T
pumpkin	3	4	1	F
strawberry	1	1	2	T

Table 7-2. Fruit Data mapped into three dimensions

There are some collisions in this data: **cherry** and **cranberry** both map to the point (1,1,1), and **banana** and **pear** both map to the point (2,2,1). If the algorithm fails to find a separating hyperplane for one such mapping, it can try another mapping. For example, instead of 1 for SMALL, 2 for MEDIUM, and 3 for LARGE, it could try 2 for SMALL, 3 for MEDIUM, and 1 for LARGE.

Here is a simple example to illustrate these ideas. Suppose the training set has two attributes, A_1 and A_2, with ranges $A_1 = \{a_{11}, a_{12}, a_{13}\}$ and $A_2 = \{a_{21}, a_{22}, a_{23}\}$. Furthermore, suppose that the training set $S = \{x, y, z\}$, where the three data points have the attribute values as shown in *Table 7-3*. The SVM algorithm will try to find a hyperplane that separates S into the two categories $C_1 = \{x, z\}$ and $C_2 = \{y\}$, according to the *Target* values of the points.

If the algorithm mappings assign 1 to a_{13}, 2 to a_{12}, 3 to a_{13}, 1 to a_{21}, 2 to a_{22}, and 3 to a_{23}, it will transform *Table 7-3* into *Table 7-4*:

Name	A1	A2	Target
x	a11	a21	T
y	a12	a22	F
z	a13	a23	T

Table 7-3. Example Training Set

Name	A1	A2	Target
x	1	1	T
y	2	2	F
z	3	3	T

Table 7-4. Mapped Attributes

Those three points are plotted in *Figure 7-12*:

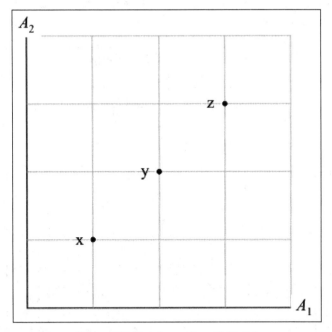

Figure 7-12. Training set from Table 7-4

In that case, there is no hyperplane that could separate the T points (**x** and **z**) from the F point (**y**) because it lies between them.

Classification Analysis

But if we renumber the values of attribute A_2, as shown in *Table 7-5*, and plot those points as shown in *Figure 7-13*, then we do have an optimal hyperplane that separates S into the two categories C_1 and C_2.

Name	A1	A2	Target
x	1	2	T
y	2	3	F
z	3	1	T

Table 7-5. Re-Mapped Attributes

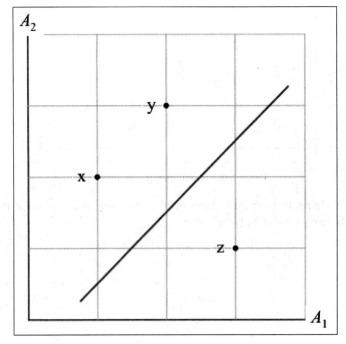

Figure 7-13. Training Set from Table 7-5

The only difficulty with this strategy is that the number of permutations of n values is $n!$. With $n = 3$ values, we have only $3! = 6$ permutations, so checking them all is no problem. But if an attribute has 30 values, then (in the worst case) there would be $30! = 2.65 \cdot 10^{32}$ different permutations to check. That is not practical. Even 10 values have $10! = 3,628,800$ permutations.

> Recall that n factorial means $n! = n \times (n-1) \times (n-2) \times \ldots \times 2 \times 1$. So, for example, $5! = 120$.

The object of the SVM algorithm is to find a hyperplane to separate all the given data points (the training set) into two given categories, which can be used to classify future data points. Ideally, we would like the hyperplane to be equidistant from the training points that are closest to it. Put another way, we want to find two parallel hyperplanes, as far apart as possible, that pass through some of the points and have none of the points between them. The distance between them is called the margin; the objective hyperplane lies midway between them in the margin. In *Figure 7-14*, the data points that lie on the bounding hyperplanes (lines) are drawn as open circles:

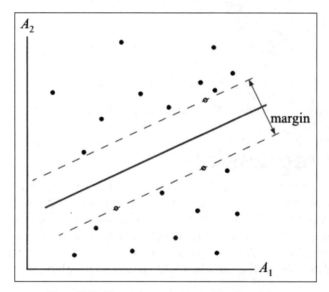

Figure 7-14. Hyperplane gap separating data points

Classification Analysis

If a (flat) hyperplane cannot be found to separate the points into the two categories, then the next approach is to seek a (curved) hypersurface to do the job. In two dimensions, that would be an ordinary curve, as shown in *Figure 7-15*:

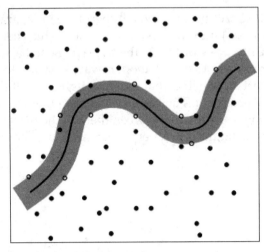

Figure 7-15. Hypersurface gap separation

Logistic regression

A classification algorithm is a process whose input is a training set, as previously described, and whose output is a function that classifies data points. The ID3 algorithm produces a decision tree for the classification function. The naive Bayes algorithm produces a function that classifies by computing ratios from the training set. The SVM algorithm produces an equation of a hyperplane (or hypersurface) that classifies a point by computing on which side of the hyperplane the point lies.

In all three of these algorithms, we assumed that all the attributes of the training set were nominal. If the attributes are instead numeric, we can apply linear regression, as we did in *Chapter 6, Regression Analysis*. The idea of logistic regression is to transform a problem whose target attribute is Boolean (that is, its value is either 0 or 1) into a numeric variable, run linear regression on that transformed problem, and then transform the solution back into the terms of the given problem. We will illustrate this process with a simple example.

Suppose some local political candidate is wondering how much money should be spent to get elected. A political research company has accumulated the data shown in *Table 7-6*, about previous elections that have similar circumstances:

Dollars Spent (× $1000)	Number of Elections	Number Who Won	Relative Frequency
1-10	6	2	2/6 = 0.3333
11-20	5	2	2/5 = 0.4000
21-30	8	4	4/8 = 0.5000
31-40	9	5	5/9 = 0.5556
41-50	5	3	3/5 = 0.6000
51-60	5	4	4/5 = 0.8000

Table 7-6. Political candidate example data

For example, among those six candidates who spent $1,000-$10,000 on their campaigns, only two won the election. The relative frequency, p is in the range $0 \le p \le 1$. We want to predict that, as a probability of winning, for other candidates whose spending x (in $1000) is given.

Instead of running linear regression on p, we run it on a dependent variable y that is a function of p. To see what that choice of function is, we first look at the odds of winning. If p is the probability of winning, then $p/(1-p)$ are the odds of winning. For example, two out of six of the candidates in the $1,000-$10,000 range won, so four of them lost; projecting those ratios as frequencies, we see that the odds of them winning are 2:4, or 2/4 = 0.5. That's $p/(1-p) = 0.3333/(1 - 0.3333) = 0.3333/0.6667 = 0.5000$.

x	p	Odds	y
5	0.3333	0.5000	-0.6931
15	0.4000	0.6667	-0.4055
25	0.5000	1.0000	0.0000
35	0.5556	1.2500	0.2231
45	0.6000	1.5000	0.4055
55	0.8000	4.0000	1.3863

Table 7-7. Converted data

Classification Analysis

The **Odds** of an event can be any number ≥ 0. To complete our complete transformation function, we take the natural logarithm of the Odds. The result could be any real number, positive or negative. This last step yields a more balanced distribution, centered at $y = 0$, which corresponds to even odds (if $p = 1/2$, then $p/(1 - p) = 1$, and $\ln 1 = 0$).

So, our transformation from p (in the range 0 to 1) to y (ranging over all real numbers) is given by the formula:

$$y = \ln\left(\frac{p}{1-p}\right)$$

This function is known as the **logit function**. Its values for our political candidate example are shown in *Table 7-7* (here x is the representing average of the interval of values. For example, 25 represents any expenditure in the range of $20,001 to $30,000).

After making this conversion from the relative frequency p to the variable y, the algorithm then runs linear regression of y on x. That requires inverting the logit function:

$$y = \ln\left(\frac{p}{1-p}\right)$$

$$\frac{p}{1-p} = e^y$$

$$\frac{1}{p} - 1 = \frac{1-p}{p} = e^{-y}$$

$$\frac{1}{p} = 1 + e^{-y}$$

$$p = \frac{1}{1+e^{-y}}$$

This is called the **sigmoid curve**. It is a special case of the more general **logistic function**, whose equation is:

$$f(x) = \frac{1}{1+e^{-k(x-x_0)}}$$

Here x_0 is the x-value of the inflection point where the slope is maximal, and k is twice that maximum slope. The curve was first studied and named by *Pierre F. Verhulst* in the 1840s, in his studies of population growth.

The graph shown in *Figure 7-16* is of the curve with the parameters $k = 1$ and $x_0 = 0$:

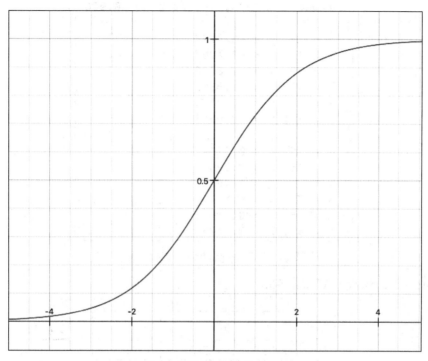

Figure 7-16. The logistic curve

You can see that it is asymptotic on the left toward $x = 0$ and on the right toward $x = 1$ and that it is symmetric about its inflection point at $(0, 0.5)$.

Classification Analysis

A Java program that computes these values is shown in *Listing 7-10*, with output in *Figure 7-17*:

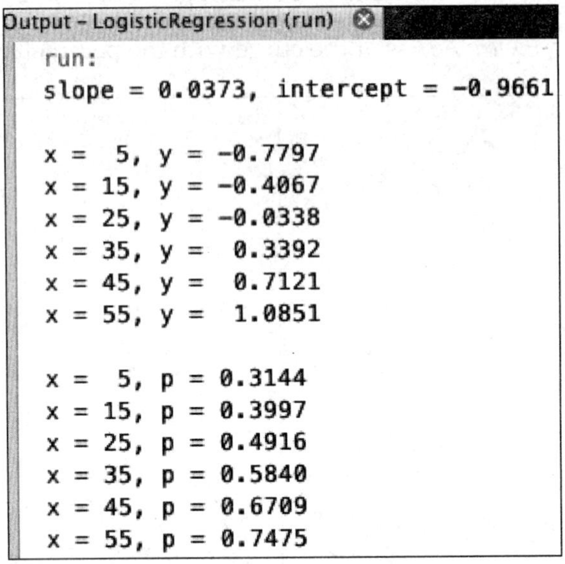

```
Output - LogisticRegression (run)
run:
slope = 0.0373, intercept = -0.9661

x =  5, y = -0.7797
x = 15, y = -0.4067
x = 25, y = -0.0338
x = 35, y =  0.3392
x = 45, y =  0.7121
x = 55, y =  1.0851

x =  5, p = 0.3144
x = 15, p = 0.3997
x = 25, p = 0.4916
x = 35, p = 0.5840
x = 45, p = 0.6709
x = 55, p = 0.7475
```

Figure 7-17. Output from Listing 7-10

The results are shown in *Table 7-8*:

x	p	y	$\hat{p}$
5	0.3333	-0.780	0.3144
15	0.4000	-0.407	0.3997
25	0.5000	0.034	0.4916
35	0.5556	0.339	0.5840
45	0.6000	0.712	0.6709
55	0.8000	1.085	0.7475

Table 7-8. Logistic regression results

The program hard codes input values at lines 12-14. Then it generates the y-values at lines 19-23. Note the use of the logit function, implemented as a class in the `commons.math3` library:

```java
 8  import org.apache.commons.math3.analysis.function.*;
 9  import org.apache.commons.math3.stat.regression.SimpleRegression;
10
11  public class LogisticRegression {
12      static int n = 6;
13      static double[] x = {5, 15, 25, 35, 45, 55};
14      static double[] p = {2./6, 2./5, 4./8, 5./9, 3./5, 4./5};
15      static double[] y = new double[n];    // y = logit(p)
16
17      public static void main(String[] args) {
18
19          // Transform p-values into y-values:
20          Logit logit = new Logit();
21          for (int i = 0; i < n; i++) {
22              y[i] = logit.value(p[i]);
23          }
24
25          // Set up input array for linear regression:
26          double[][] data = new double[n][n];
27          for (int i = 0; i < n; i++) {
28              data[i][0] = x[i];
29              data[i][1] = y[i];
30          }
31
32          // Run linear regression of y on x:
33          SimpleRegression sr = new SimpleRegression();
34          sr.addData(data);
35
36          // Print results:
37          System.out.printf("slope = %.4f, intercept = %.4f%n%n",
38                  sr.getSlope(), sr.getIntercept());
39          for (int i = 0; i < n; i++) {
40              System.out.printf("x = %2.0f, y = %7.4f%n", x[i], sr.predict(x[i]));
41          }
42          System.out.println();
43
44          // Convert y-values back to p-values:
45          Sigmoid sigmoid = new Sigmoid();
46          for (int i = 0; i < n; i++) {
47              double p = sr.predict(x[i]);
48              System.out.printf("x = %2.0f, p = %6.4f%n", x[i], sigmoid.value(p));
49          }
50      }
51  }
```

Listing 7-10. Logistic regression example

Classification Analysis

To use the `SimpleRegression` object from that library, we first set up the x and y data in a two-dimensional array at lines 25-30. It then gets added into that object at line 34. At line 37-38, we get from the `sr` object and print the slope and y-intercept of the regression line of y on x. Then we use its `predict()` method at line 40 to print the y-values. Then finally, at line 48, we use a sigmoid function to compute and print the corresponding p-values.

The equation of the regression line for this training set is:

$$y = 0.0373x - 0.9661$$

Therefore, the equation of the logistic regression curve for this data is:

$$p = \frac{1}{1 + e^{-0.0373x + 0.9661}}$$

We can use that to estimate the probability that a candidate will win a local election after spending x on the campaign. For example, an expenditure of $48,000 (x = 48) would suggest this probability of winning:

$$p = \frac{1}{1 + e^{-0.0373(48) + 0.9661}} = \frac{1}{1 + e^{-0.824}} = \frac{1}{1 + 0.439} = 0.695$$

K-Nearest Neighbors

Nearest neighbor algorithms are some of the simplest classification methods available. All you need is some kind of metric (distance function) and the assumption that similar points tend to be near each other in the attribute space.

The k in the name of this algorithm (abbreviated k-NN) stands for an assumed constant. Its value is used as a threshold number of neighbors, whose value (category) of the target attribute is known. The most common value among them is assigned to the new point.

As an example of how the k-NN algorithm works, look at the data points in *Figure 7-18*:

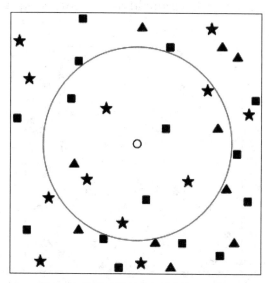

Figure 7-18. K-Nearest Neighbor example

The (visible part of the) training set has 37 points, with three attributes: x-coordinate, y-coordinate, and shape. The shape attribute, which is the target, has three possible values (categories): triangle, square, and star. Of the 37 points, 10 are triangles, 15 are squares, and 12 are stars. The algorithm is to predict the category of the new unclassified point, shown as a tiny circle.

The first step of the algorithm is to assign a number to the constant k. We will choose $k = 5$, which means that the algorithm will continue until it has found five points of the same category. It begins with a circle of radius $r = 0$, centered at the new point, and begins expanding the radius, keeping count of how many data points of each type are within the circle. The diagram in *Figure 7-18* shows the situation just after the circle has encompassed the fifth star. Since $k = 5$ is the threshold, the expansion stops. The first category to have five representatives within the expanding circle is the star. There are only three squares and only two triangles inside the circle at that point. So, the algorithm predicts that the new point is a star.

Classification Analysis

The choice of the value of the constant k may seem to have an unstable effect upon the outcome of the algorithm. For example, if we had chosen $k = 2$ instead of 5, *Figure 7-19* shows that the algorithm would have predicted square instead of star for the new point:

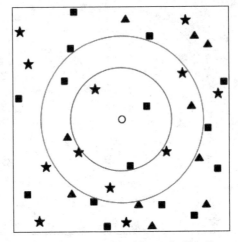

Figure 7-19. Using k =2 Instead of k = 5

Of course, this particular training set shows a severe lack of affinity among similar points.

We can compute the accuracy of a specific choice of k by running the algorithm with that k on each point in the training set and simply counting how many of those get predicted correctly. *Figure 7-20* shows the testing of two of the 37 training set points, using $k = 2$:

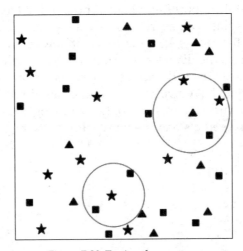

Figure 7-20. Testing the accuracy

Chapter 7

> Both predictions are wrong: the triangle is predicted to be a star, and the star is predicted to be a square. This training set is clearly not very useful.

The Java program in *Listing 7-11* tests the basic k-NN algorithm with the default value of *k* = 1. Note that Weka's implementation of this algorithm is their `IBk` class:

```java
import weka.classifiers.lazy.IBk;   // K-Nearest Neighbors
import weka.core.Instances;
import weka.core.Instance;
import weka.core.converters.ConverterUtils.DataSource;

public class TestIBk {
    public static void main(String[] args) throws Exception {
        DataSource source = new DataSource("data/AnonFruit.arff");
        Instances instances = source.getDataSet();
        instances.setClassIndex(3);   // target attribute: (Sweet)

        IBk ibk = new IBk();
        ibk.buildClassifier(instances);

        for (Instance instance : instances) {
            double prediction = ibk.classifyInstance(instance);
            System.out.printf("%4.0f%4.0f%n",
                    instance.classValue(), prediction);
        }
    }
}
```

```
run:
   1   1
   1   1
   1   1
   1   0
   1   1
   0   0
   1   1
   0   0
   0   0
   0   0
   1   1
   1   1
   1   1
   1   1
   0   0
   1   1
```

Listing 7-11. K-NN implementation using Weka's IBk class

Classification Analysis

This is similar to the naive Bayes algorithm test in *Listing 7-8*.

The accuracy of the choice of k can be defined as the percentage of correct predictions for the points in the training set. If that percentage is low, we can attempt better accuracy by adjusting the value of k.

Note that in this example, we have only two dimensions (only two attributes, not counting the target attribute). A more realistic example might have 20 dimensions. In higher-dimensional classification, we can also attempt to improve the accuracy by eliminating some of the dimensions. That means, just ignore one or more of the attributes.

There are several other strategies for improving the accuracy of the basic k-NN algorithm. Some of these weight the attributes with a weight vector $\mathbf{w} = (w_1, w_2,..., w_n)$, where the individual weight numbers w_i are specified in advance and then adjusted to improve accuracy. This is equivalent to redefining the distance function as:

$$d(\mathbf{x}, \mathbf{y}) = \sum_{i=1}^{n} w_i (x_i - y_i)^2$$

In fact, the previously mentioned strategy of ignoring an attribute can be seen as a special case of this weighting strategy. For example, if we have $n = 6$ attributes and we decide to drop the third attribute, we would simply define the weight vector as $\mathbf{w} = (1, 1, 0, 1, 1, 1)$.

Any classification algorithm itself is dependent upon the training set that initializes it. If we modify the training test, we are modifying the classification algorithm that is built upon it. Some classification algorithms depend upon a parameter that can be easily changed. The k-NN algorithm depends upon the parameter k.

Here is an algorithm, called **backward elimination**, for managing this strategy. The accuracy α is defined as the percentage of correct predictions made on the training set:

1. Compute a for the training set using all the attributes.
2. Repeat, for $i = 1$ to n:
    1. Delete attribute A_i.
    2. Compute a'.
    3. If $a' < a$, restore attribute A_i.

The k-NN algorithm is a classification process that has many important applications, including:

1. Facial recognition.
2. Text classification.
3. Gene expression.
4. Recommendation systems.

Fuzzy classification algorithms

The general idea of fuzzy mathematics is to replace set membership with probability distribution. If, in some algorithm, we replace the Boolean statement $x \in A$ with the probabilistic statement $P(x > x_0)$, we have transformed it into a fuzzy algorithm.

The purpose of a classification algorithm is to predict to which category a data point belongs. Is this fruit sweet? Will that politician win the election? What color is this data point? With a fuzzy classification algorithm, we instead answer, "What is the probability distribution for this data point relative to the various possible outcomes?"

Instead of requiring every data point to belong to a single category of the target attribute, we assume every point has a probability vector p that specifies the likelihood that the point belongs to each category:

$$\mathbf{p} = (p_1, p_2, \cdots, p_n), \forall p_i \geq 0, \sum p_i = 1$$

For example, in our star-square-triangle example, the vector $p = (1/3, 2/3, 0)$ would indicate that the point is either a star or a square, and is twice as likely to be a square. Or, if we're thinking about data that is somehow fluid, changing from one state to another, then that p would indicate that the point is always either a star or a square, and that it's a square twice as often as it is a star. That model might be appropriate, for example, in an application that attempts to classify voters by political orientation, such as liberal, conservative, or libertarian.

Summary

This chapter presents some of the main algorithms that have been developed in recent years to classify data: decision trees, ID3, Naïve Bayes, SVM, logistic regression, and k-NN. We have shown how to implement some of these in Java, using the Weka and the Apache Commons Math libraries.

8
Cluster Analysis

A clustering algorithm is one that identifies groups of data points according to their proximity to each other. These algorithms are similar to classification algorithms in that they also partition a dataset into subsets of similar points. But, in classification, we already have data whose classes have been identified. such as sweet fruit. In clustering, we seek to discover the unknown groups themselves.

Measuring distances

A **metric** on a set S of points is a function $d : S \times S \to \mathbb{R}$ that satisfies these conditions for all $x, y, z \in S$:

1. $d(p,q) = 0 \Leftrightarrow p=q$
2. $d(p,q) = d(p,q)$
3. $d(p,q) \leq d(p,r) + d(r,q)$

Normally, we think of the number $d(p,q)$ as the distance between p and q. With that interpretation, the three conditions are obvious: the distance from a point to itself is 0; if the distance between two points is 0, then they must be the same point; the distance from p to q is the same as the distance from q to p; the distance from p to q cannot be greater than the sum of the distances from p to r and from r to q. This last property is called the **triangle inequality**.

In mathematics, a non-empty set S together with a metric d defined on it is called a **metric space**. The simplest example is n-dimensional Euclidean space $(\mathbb{R}^n, d)$, where $\mathbb{R}^n = \{(x_1, \cdots, x_n) : x_j \in \mathbb{R}\}$ and d is the Euclidean metric,

$$d(\mathbf{x}, \mathbf{y}) = \sqrt{\sum_{j=1}^{n} (x_j - y_j)^2}$$

Cluster Analysis

In the case of two dimensions, $\mathbb{R}^2 = \{(x_1, x_2) : x_1, x_2 \in \mathbb{R}\}$, and $d(\mathbf{x},\mathbf{y}) = \sqrt{(x_1-y_1)^2 + (x_2-y_2)^2}$. This is just the ordinary distance formula for points in the **Cartesian plane**, equivalent to the **Pythagorean theorem**, as *Figure 8.1* shows:

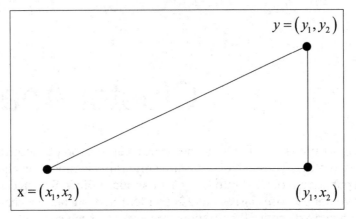

Figure 8.1: Euclidean metric in two dimensions

For example, if $x = (1, 3)$ and $y = (5, 6)$, then $d(\mathbf{x},\mathbf{y}) = \sqrt{(1-5)^2 + (3-6)^2} = \sqrt{16+9} = 5$.

In three dimensions, $\mathbb{R}^3 = \{(x_1, x_2, x_3) : x_1, x_2, x_3 \in \mathbb{R}\}$, and $d(\mathbf{x},\mathbf{y}) = \sqrt{(x_1-y_1)^2 + (x_2-y_2)^2 + (x_3-y_3)^2}$.

This is illustrated in *Figure 8.2*:

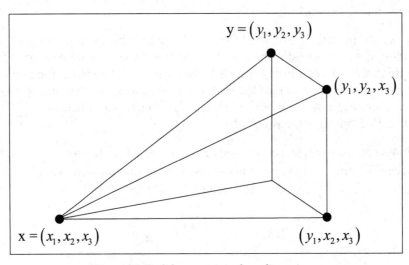

Figure 8.2: Euclidean metric in three dimensions

The **Minkowski metric** for $\mathbb{R}^n$ generalizes the Euclidean metric with a parameter $p \geq 1$:

$$d(\mathbf{x},\mathbf{y}) = \sqrt[p]{\sum_j |x_j - y_j|^p}$$

(Mathematicians call this the **L^p metric**.)

For example, with $p = 4$:

$$d(\mathbf{x},\mathbf{y}) = \sqrt[4]{\sum_j |x_j - y_j|^4}$$

That would measure the distance between **x** = (1, 3) and **y** = (5, 6) as:

$$d(\mathbf{x},\mathbf{y}) = \sqrt[4]{|1-5|^4 + |3-6|^4} = \sqrt[4]{337} = 4.28$$

The most important special case of the Minkowski metric (after the case of $p = 2$) is the case of $p = 1$:

$$d(\mathbf{x},\mathbf{y}) = \sqrt[1]{\sum_j |x_j - y_j|^1} = \sum_j |x_j - y_j|$$

In two dimensions, this would be:

$$d(\mathbf{x},\mathbf{y}) = \sum_j |x_j - y_j| = |x_1 - y_1| + |x_2 - y_2|$$

For example, if $x = (1, 3)$ and $y = (5, 6)$, then $d(\mathbf{x},\mathbf{y}) = |1-5| + |3-6| = 4 + 3 = 7$.

Cluster Analysis

This special case is also called the **Manhattan metric** (or the **taxicab metric**), because in two dimensions it models the distance travelled on a grid, such as the streets of Manhattan, as illustrated in *Figure 8.3*:

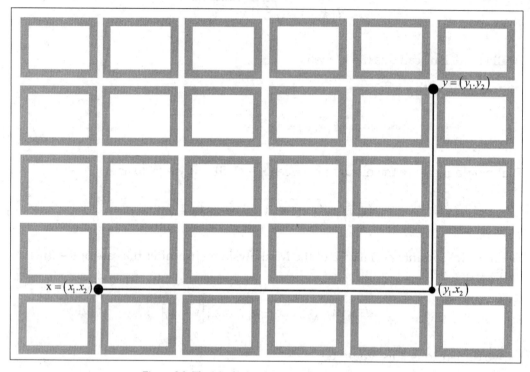

Figure 8.3: The Manhattan Metric in Two Dimensions

Mathematically, if we take the limit as p approaches infinity, the Minkowski metric converges to the **Chebyshev metric**:

$$d(\mathbf{x},\mathbf{y}) = \max\left\{\left|x_j - y_j\right|\right\}$$

In two dimensions, this is:

$$d(\mathbf{x},\mathbf{y}) = \max\left\{\left|x_1 - y_1\right|, \left|x_2 - y_2\right|\right\}$$

For example, if x = (1, 3) and y = (5, 6), then:

$$d(\mathbf{x},\mathbf{y}) = \max\left\{\left|1-5\right|, \left|3-6\right|\right\} = \max\{4,3\} = 4$$

This is also called the **chessboard metric**, because it is the correct number of moves a king must make to go from one square to another on a chessboard.

Another specialized metric that is useful in machine learning is the **Canberra metric**:

$$d(\mathbf{x},\mathbf{y}) = \sum_{j=1}^{n} \frac{|x_j - y_j|}{|x_j| + |y_j|}$$

This is a weighted version of the Manhattan metric.

For example, if $x = (1, 3)$ and $y = (5, 6)$, then:

$$d(\mathbf{x},\mathbf{y}) = \sum_{j=1}^{2} \frac{|x_j - y_j|}{|x_j| + |y_j|} = \frac{|1-5|}{|1|+|5|} + \frac{|3-6|}{|3|+|6|} = \frac{4}{6} + \frac{3}{9} = 1$$

Note that the Canberra metric is bounded by the dimension n. This follows from the triangle inequality:

$$|x_j - y_j| = |x_j + (-y_j)| \leq |x_j| + |-y_j| = |x_j| + |y_j|$$

$$\frac{|x_j - y_j|}{|x_j| + |y_j|} \leq 1$$

$$d(\mathbf{x},\mathbf{y}) = \sum_{j=1}^{n} \frac{|x_j - y_j|}{|x_j| + |y_j|} \leq \sum_{j=1}^{n} 1 = n$$

For example, in the Cartesian plane, $\mathbb{R}^2$, all Canberra distances are less and/or equal to two. Even more counter-intuitive is the fact that every nonzero point is the same distance from the origin 0:

$$d(\mathbf{x},\mathbf{0}) = \sum_{j=1}^{n} \frac{|x_j - 0|}{|x_j| + |0|} = \sum_{j=1}^{n} \frac{|x_j|}{|x_j|} = \sum_{j=1}^{n} 1 = n$$

Cluster Analysis

The program in *Listing 8.1* computes these examples using the corresponding methods that are defined in the `org.apache.commons.math3.ml.distance` package.

```java
import org.apache.commons.math3.ml.distance.*;

public class TestMetrics {
    public static void main(String[] args) {
        double[] x = {1, 3}, y = {5, 6};

        EuclideanDistance eD = new EuclideanDistance();
        System.out.printf("Euclidean distance = %.2f%n", eD.compute(x,y));

        ManhattanDistance mD = new ManhattanDistance();
        System.out.printf("Manhattan distance = %.2f%n", mD.compute(x,y));

        ChebyshevDistance cD = new ChebyshevDistance();
        System.out.printf("Chebyshev distance = %.2f%n", cD.compute(x,y));

        CanberraDistance caD = new CanberraDistance();
        System.out.printf("Canberra distance = %.2f%n", caD.compute(x,y));
    }
}
```

```
run:
Euclidean distance = 5.00
Manhattan distance = 7.00
Chebyshev distance = 4.00
Canberra distance = 1.00
```

Listing 8.1: Testing metrics in the commons math library

Finally, we see in *Figure 8.4* how metrics can be compared geometrically. It shows the two-dimensional unit ball:

$$B_2 = \left\{ \mathbf{x} \in \mathbb{R}^2 : d(\mathbf{x},0) \leq 1 \right\}$$

It does this for each of the three metrics: Euclidean, Manhattan, and Chebyshev:

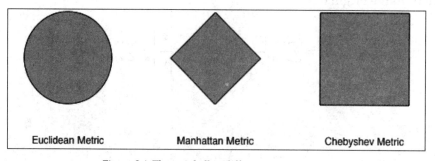

Figure 8.4: The unit ball in different metric spaces

The curse of dimensionality

Most clustering algorithms depend upon the distances between points in the data space. But it is a fact of Euclidean geometry that average distances grow as the number of dimensions increases.

For example, look at the unit hypercube:

$$H_n = \{\mathbf{x} \in \mathbb{R}^n : 0 \leq x_j \leq 1 \text{ for } j = 1, \cdots, n\}$$

The one-dimensional hypercube is the unit interval *[0,1]*. The two points that are farthest apart in this set are *0* and *1*, whose distance *d(0,1) = 1*.

The two-dimensional hypercube is the unit square. The two points that are farthest apart in H_2 are the corner points *0 = (0,0)* and *x = (1,1)*, whose distance is $d(\mathbf{0},\mathbf{x}) = \sqrt{2}$.

In H_n, the two corner points *0 = (0, 0, ..., 0)* and *x = (1, 1, ..., 1)* are at the distance $d(\mathbf{0},\mathbf{x}) = \sqrt{n}$.

Not only do points tend to be farther apart in higher-dimensional space, but also their vectors tend to be perpendicular. To see that, suppose $\mathbf{x} = (x_1, \ldots, x_n)$ and $\mathbf{y} = (y_1, \ldots, y_n)$ are points in $\mathbb{R}^n$. Recall that their dot product (also called the scalar product) is $\mathbf{x} \cdot \mathbf{y} = \sum_j x_j y_j = x_1 y_1 + \cdots x_n y_n$. But we also have this formula from the Law of Cosines: $\mathbf{x} \cdot \mathbf{y} = |\mathbf{x}||\mathbf{y}|\cos\theta$, where θ is the angle between the two vectors.

For simplicity, suppose also that $|\mathbf{x}| = |\mathbf{y}| = 1$ (that is, **x** and **y** are unit vectors). Then, by combining these formulas, we have:

$$\cos\theta = \sum_{j=1}^{n} x_j y_j$$

Cluster Analysis

Now, if the number of dimensions is large, say $n = 100$, and if x and y are unit vectors, then their components, x_j and y_j must be small, which means that the sum on the right must be small and that means that that θ must be near 90°; that is, the vectors are nearly perpendicular.

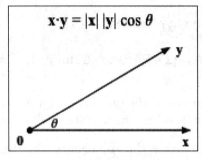

Figure 8.5: Dot product

If x and y are not unit vectors, the formula is:

$$\cos\theta = \frac{\sum_{j=1}^{n} x_j y_j}{|x||y|}$$

This quotient is likely to be near zero if n is large, giving us the same conclusion as before: that x and y are nearly perpendicular.

For example, suppose that $n = 100$, $\mathbf{x} = (-1,1,-1,1,\ldots,1)$, and $\mathbf{y} = (1,2,3,4,\ldots,100)$. Then:

$$|\mathbf{x}|^2 = \sum_j \left((-1)^j\right)^2 = \sum_j (-1)^{2j} = \sum_j 1 = 100 \Rightarrow |\mathbf{x}| = 10$$

$$|\mathbf{y}|^2 \sum_j j^2 = 338{,}350 \Rightarrow |\mathbf{y}| = \sqrt{338{,}350} = 581.7$$

$$\mathbf{x}\cdot\mathbf{y} = \sum_j x_j y_j = \sum_j (-1)^j j = -1+2-3+4-\cdots+100 = 50$$

$$\cos\theta = \frac{\mathbf{x}\cdot\mathbf{y}}{|\mathbf{x}||\mathbf{y}|} = \frac{50}{(10)(581.7)} = 0.008596 \Rightarrow \theta = 89.5°$$

You can see from this example the reason that **x** · **y** in general is likely to be much smaller than $|x||y|$: cancelation. On average, we would expect about half of the $x_j y_j$ terms to be negative.

So, in high dimensions, most vectors are nearly perpendicular. That means that the distance between two points x and y will be greater than the distance of either from the origin. Algebraically, $x \cdot y \approx 0$, so:

$$|\mathbf{x}-\mathbf{y}|^2 = (\mathbf{x}-\mathbf{y})\cdot(\mathbf{x}-\mathbf{y}) = \mathbf{x}\cdot\mathbf{x} - 2\mathbf{x}\cdot\mathbf{y} + \mathbf{y}\cdot\mathbf{y} \approx \mathbf{x}\cdot\mathbf{x} + \mathbf{y}\cdot\mathbf{y} = |\mathbf{x}|^2 + |\mathbf{y}|^2 \geq |\mathbf{x}|^2$$

Thus the distance $|x - y|$ will be larger than $|x|$, and larger than $|y|$ for the same reason. This is called the **curse of dimensionality**.

Hierarchical clustering

Of the several clustering algorithms that we will examine in this chapter, hierarchical clustering is probably the simplest. The trade-off is that it works well only with small datasets in Euclidean space.

The general setup is that we have a dataset S of m points in $\mathbb{R}^n$ which we want to partition into a given number k of clusters $C_1, C_2, ..., C_k$, where within each cluster the points are relatively close together. (B. J. Frey and D. Dueck, *Clustering by Passing Messages Between Data Points* Science 315, Feb 16, 2007 http://science.sciencemag.org/content/315/5814/972).

Here is the algorithm:

1. Create a singleton cluster for each of the *m* data points.
2. Repeat *m* – *k* times:
    - Find the two clusters whose centroids are closest
    - Replace those two clusters with a new cluster that contains their points

The centroid of a cluster is the point whose coordinates are the averages of the corresponding coordinates of the cluster points. For example, the centroid of the cluster C = {(2, 4), (3, 5), (6, 6), (9, 1)} is the point **(5, 4)**, because (2 + 3 + 6 + 9)/4 = 5 and (4 + 5 + 6 + 1)/4 = 4. This is illustrated in *Figure 8.6*:

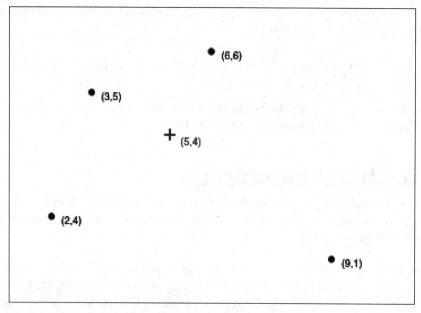

Figure 8.6: The centroid of a cluster

A Java implementation of this algorithm is shown in *Listing 8.2*, with partial output in *Figure 8.7*. It uses a `Point` class and a `Cluster` class, which are shown in *Listing 8.3* and *Listing 8.4*, respectively.

The dataset is defined at lines 11-12 in *Listing 8.2*:

```java
public class HierarchicalClustering {
    private static final double[][] DATA = {{1,1}, {1,3}, {1,5}, {2,6}, {3,2},
        {3,4}, {4,3}, {5,6}, {6,3}, {6,4}, {7,1}, {7,5}, {7,6}};
    private static final int M = DATA.length;   // number of points
    private static final int K = 3;             // number of clusters

    public static void main(String[] args) {
        HashSet<Cluster> clusters = load(DATA);
        for (int i = 0; i < M - K; i++) {
            System.out.printf("%n%2d clusters:%n", M-i-1);
            coalesce(clusters);
            System.out.println(clusters);
        }
    }

    private static HashSet<Cluster> load(double[][] data) {
        HashSet<Cluster> clusters = new HashSet();
        for (double[] datum : DATA) {
            clusters.add(new Cluster(datum[0], datum[1]));
        }
        return clusters;
    }

    private static void coalesce(HashSet<Cluster> clusters) {
        Cluster cluster1=null, cluster2=null;
        double minDist = Double.POSITIVE_INFINITY;
        for (Cluster c1 : clusters) {
            for (Cluster c2 : clusters) {
                if (!c1.equals(c2) && Cluster.distance(c1, c2) < minDist) {
                    cluster1 = c1;
                    cluster2 = c2;
                    minDist = Cluster.distance(c1, c2);
                }
            }
        }
        clusters.remove(cluster1);
        clusters.remove(cluster2);
        clusters.add(Cluster.union(cluster1, cluster2));
    }
}
```

Listing 8.2. An implementation of hierarchical clustering

These 13 data points are loaded in the `clusters` set at line 17. Then the loop at lines 18-22 iterates $m - k$ times, as specified in the algorithm.

A cluster, as defined by the `Cluster` class in *Listing 8.4*, contains two objects: a set of points and a centroid, which is a single point. The distance between two clusters is defined as the Euclidean distance between their centroids. Note that, for simplicity, some of the code in *Listing 8.4* has been folded.

Cluster Analysis

The program uses the `HashSet<Cluster>` class to implement the set of clusters. That is why the `Cluster` class overrides the `hashCode()` and `equals()` methods (at lines 52-68 in *Listing 8.4*). That, in turn, requires the `Point` class to override its corresponding methods (at lines 27-43 in *Listing 8.3*).

Note that the `Point` class defines private fields `xb` and `yb` of type `long`. These hold the 64 bit representations of the `double` values of *x* and *y*, providing a more faithful way to determine when they are equal.

The output shown in *Figure 8.7* is generated by the code at line 19 and 21 of the program. The call to `println()` at line 21 implicitly invokes the overridden `toString()` method at lines 70-74 of the `Cluster` class.

The `coalesce()` method at lines 33-49 implements the two parts of *step 2* of the algorithm. The double loop at lines 36-44 finds the two clusters that are closest to each other (*step 2* first part). These are removed from the `clusters` set and their union is added to it at lines 45-47 (*step 2*, second part).

The output in *Figure 8.7* shows the results of two iterations of the double loop: six clusters coalescing into five, and then into four.

```
Output - Clustering (run)
    6 clusters:
    [
    {(3.33,3.00),[(3.00,4.00), (3.00,2.00), (4.00,3.00)]},
    {(1.50,5.50),[(1.00,5.00), (2.00,6.00)]},
    {(6.00,3.50),[(6.00,3.00), (6.00,4.00)]},
    {(1.00,2.00),[(1.00,1.00), (1.00,3.00)]},
    {(6.33,5.67),[(5.00,6.00), (7.00,6.00), (7.00,5.00)]},
    {(7.00,1.00),[(7.00,1.00)]}]

    5 clusters:
    [
    {(3.33,3.00),[(3.00,4.00), (3.00,2.00), (4.00,3.00)]},
    {(1.50,5.50),[(1.00,5.00), (2.00,6.00)]},
    {(6.20,4.80),[(6.00,3.00), (6.00,4.00), (5.00,6.00), (7.00,6.00), (7.00,5.00)]},
    {(1.00,2.00),[(1.00,1.00), (1.00,3.00)]},
    {(7.00,1.00),[(7.00,1.00)]}]

    4 clusters:
    [
    {(1.50,5.50),[(1.00,5.00), (2.00,6.00)]},
    {(6.20,4.80),[(6.00,3.00), (6.00,4.00), (5.00,6.00), (7.00,6.00), (7.00,5.00)]},
    {(7.00,1.00),[(7.00,1.00)]},
    {(2.40,2.60),[(3.00,4.00), (3.00,2.00), (4.00,3.00), (1.00,1.00), (1.00,3.00)]}]
```

Figure 8.7: Partial output from hierarchical clustering program

Chapter 8

These three stages are illustrated in *Figure 8.8*, *Figure 8.9*, and *Figure 8.10*.

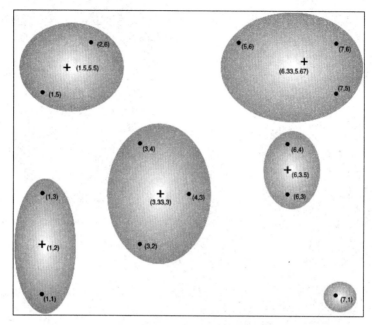

Figure 8.8: Output with six clusters

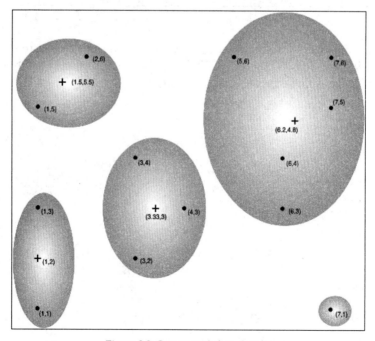

Figure 8.9: Output with five clusters

Cluster Analysis

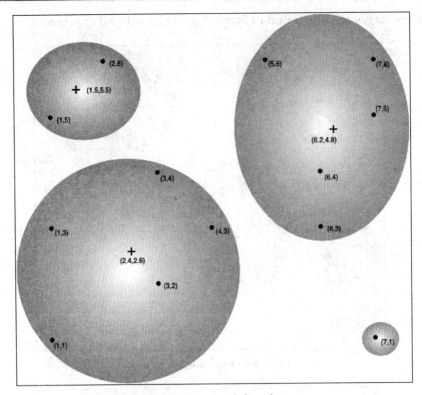

Figure 8.10: Output with four clusters

Among the six clusters shown in *Figure 8.8*, you can see the closest are the ones whose centroids are (6.00, 3.50) and (6.33, 5.67). They get replaced by their union, the five-element cluster whose centroid is (6.20, 4.80).

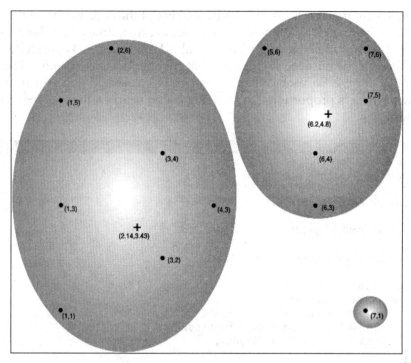

Figure 8.11: Output with three clusters

We can set the number K to any value from 1 to M (at line 14). Although setting K = 1 has no practical value, it can be used to generate the **dendrogram** for the original dataset, shown in *Figure 8.12*. It graphically displays the hierarchical structure of the entire clustering process, identifying each coalesce step.

Note that the dendrogram is not a complete transcript of the process. It shows, for example, that **(3,4)** and **(4,3)** get united before they unite with **(3,2)**, but it does not show whether **(1,5)** and **(2,6)** get united before or after **(1,1)** and **(1,3)** do.

Although easy to understand and implement, hierarchical clustering is not very efficient. The basic version, shown in *Listing 8.2*, runs in $O(n^3)$ time, which means that the number of operations performed is roughly proportional to the cube of the number of points in the dataset. So, if you double the number of points, the program will take about eight times as long to run. That is not practical for large datasets.

Cluster Analysis

You can see where the $O(n^3)$ comes from by looking at the code in *Listing 8.2*. There, $n = 13$. The main loop (lines 18-22) iterates nearly n times. Each iteration calls the `coalesce()` method, which has a double loop (lines 36-44), each iterating c times, where c is the number of clusters. That number decreases from n down to k, averaging about $n/2$. So, each call to the `coalesce()` method will execute about $(n/2)^2$ times, which is proportional to n^2. Being called nearly n times gives us the $O(n^3)$ runtime.

```java
public class Point {
    private final double x, y;

    public Point(double x, double y) {...4 lines }

    public double getX() {...3 lines }

    public double getY() {...3 lines }

    @Override
    public int hashCode() {
        int xhC = new Double(x).hashCode();
        int yhC = new Double(y).hashCode();
        return (int)(xhC + 79*yhC);
    }

    @Override
    public boolean equals(Object object) {
        if (object == null) {
            return false;
        } else if (object == this) {
            return true;
        } else if (!(object instanceof Point)) {
            return false;
        }
        Point that = (Point)object;
        return bits(that.x) == bits(this.x) && bits(that.y) == bits(this.y);
    }

    private long bits(double d) {
        return Double.doubleToLongBits(d);
    }

    @Override
    public String toString() {
        return String.format("(%.2f,%.2f)", x,y);
    }
}
```

Listing 8.3: A Point class

This kind of **complexity analysis** is standard for computer algorithms. The main idea is to find some simple function $f(n)$ that classifies the algorithm this way. The $O(n^3)$ classification means slow. (The letter O stands for "order of". So, $O(n^3)$ means "on the order of n^3".)

We can bump up the runtime classification of the hierarchical clustering algorithm by using a more elaborate data structure. The idea is to maintain a priority queue (a balanced binary tree structure) of objects, where each object consists of a pair of points and the distance between them. Objects can be inserted into and removed from a priority queue in $O(\log n)$ time. Consequently, the whole algorithm can be run in $O(n^2 \log n)$, which is almost as good as $O(n^2)$.

This improvement is a good example of the classic alternative of speed vs. simplicity in computing: often, we can make an algorithm faster (more efficient) at the cost of making it more complicated. Of course, the same thing can be said about cars and airplanes.

The `Point` class in *Listing 8.3* encapsulates the idea of a two-dimensional point in Euclidean space. The `hashCode()` and `equals()` methods must be included (overriding the default versions defined in the `Object` class) because we intend to use this class as the element type in a `HashSet` (in *Listing 8.4*).

The code at lines 26-29 defines a typical implementation; the expression `new Double(x).hashCode()` at line 26 returns the `hashCode` of the `Double` object that represents the value of x.

The code at lines 33-42 similarly defines a typical implementation of an `equals()` method. The first statement (lines 33-39) checks whether the explicit object is `null`, equals the implicit object (`this`), and is itself an instance of the `Point` class, taking the appropriate action in each case. If it passes those three tests, then it is recast as a `Point` object at line 40 so that we can access its x and y fields.

Cluster Analysis

To check whether they match the corresponding type `double` fields of the implicit object, we use an auxiliary `bits()` method, which simply returns a `long` integer containing all the bits that represent the specified `double` value.

```java
public class Cluster {
    private final HashSet<Point> points;
    private Point centroid;

    public Cluster(HashSet points, Point centroid) {...4 lines}

    public Cluster(Point point) {...5 lines}

    public Cluster(double x, double y) {...3 lines}

    public Point getCentroid() {
        return centroid;
    }

    public void add(Point point) {
        points.add(point);
        recomputeCentroid();
    }

    public void recomputeCentroid() {
        double xSum=0.0, ySum=0.0;
        for (Point point : points) {
            xSum += point.getX();
            ySum += point.getY();
        }
        centroid = new Point(xSum/points.size(), ySum/points.size());
    }

    public static double distance(Cluster c1, Cluster c2) {
        double dx = c1.centroid.getX() - c2.centroid.getX();
        double dy = c1.centroid.getY() - c2.centroid.getY();
        return Math.sqrt(dx*dx + dy*dy);
    }

    public static Cluster union(Cluster c1, Cluster c2) {
        Cluster cluster = new Cluster(c1.points, c1.centroid);
        cluster.points.addAll(c2.points);
        cluster.recomputeCentroid();
        return cluster;
    }

    @Override
    public int hashCode() {...3 lines}

    @Override
    public boolean equals(Object object) {...11 lines}

    @Override
    public String toString() {...3 lines}
}
```

Listing 8.4: A Cluster class

Weka implementation

The program in *Listing 8.5* is equivalent to that in *Listing 8.2*:

```java
import java.util.ArrayList;
import weka.clusterers.HierarchicalClusterer;
import static weka.clusterers.HierarchicalClusterer.TAGS_LINK_TYPE;
import weka.core.Attribute;
import weka.core.Instance;
import weka.core.Instances;
import weka.core.SelectedTag;
import weka.core.SparseInstance;

public class HierarchicalClustering {
    private static final double[][] DATA = {{1,1}, {1,3}, {1,5}, {2,6}, {3,2},
            {3,4}, {4,3}, {5,6}, {6,3}, {6,4}, {7,1}, {7,5}, {7,6}};
    private static final int M = DATA.length;  // number of points
    private static final int K = 3;            // number of clusters

    public static void main(String[] args) {
        Instances dataset = load(DATA);
        HierarchicalClusterer hc = new HierarchicalClusterer();
        hc.setLinkType(new SelectedTag(4, TAGS_LINK_TYPE));  // CENTROID
        hc.setNumClusters(3);
        try {
            hc.buildClusterer(dataset);
            for (Instance instance : dataset) {
                System.out.printf("(%.0f,%.0f): %s%n",
                        instance.value(0), instance.value(1),
                        hc.clusterInstance(instance));
            }
        } catch (Exception e) {
            System.err.println(e);
        }
    }

    private static Instances load(double[][] data) {
        ArrayList<Attribute> attributes = new ArrayList<Attribute>();
        attributes.add(new Attribute("X"));
        attributes.add(new Attribute("Y"));
        Instances dataset = new Instances("Dataset", attributes, M);
        for (double[] datum : data) {
            Instance instance = new SparseInstance(2);
            instance.setValue(0, datum[0]);
            instance.setValue(1, datum[1]);
            dataset.add(instance);
        }
        return dataset;
    }
}
```

```
run:
(1,1): 0
(1,3): 0
(1,5): 0
(2,6): 0
(3,2): 0
(3,4): 0
(4,3): 0
(5,6): 1
(6,3): 1
(6,4): 1
(7,1): 2
(7,5): 1
(7,6): 1
```

Listing 8.5: Hierarchical clustering with Weka

Cluster Analysis

You can see that the results are the same as illustrated in *Figure 8.11*: The first seven points are in cluster number *0*, and all the others except (7,1) are in cluster number *1*.

The `load()` method at lines 40-52 uses an `ArrayList` to specify the two attributes x and y. Then it creates the `dataset` as an `Instances` object at line 44, loads the *13* data points as `Instance` objects in the loop at lines 45-50, and returns it back to line 24.

The code at lines 25-26 specifies that the centroids of the clusters are to be used for computing the distances between the clusters.

The algorithm itself is run by the `buildClusterer()` method at line 29. Then, the loop at lines 30-34 prints the results. The `clusterInstance()` method returns the number of the cluster to which the specified `instance` belongs.

The dendrogram shown in *Figure 8.12* can be generated programmatically. You can do it with these three changes to the program in *Listing 8.5*:

1. Set the number of clusters to `1` at line 31.
2. Insert this line of code at line 39: `displayDendrogram(hc.graph());`
3. Insert the `displayDendrogram()` method shown in *Listing 8.6*:

```
59    public static void displayDendrogram(String graph) {
60        JFrame frame = new JFrame("Dendrogram");
61        frame.setSize(500, 400);
62        frame.setDefaultCloseOperation(JFrame.EXIT_ON_CLOSE);
63        Container pane = frame.getContentPane();
64        pane.setLayout(new BorderLayout());
65        pane.add(new HierarchyVisualizer(graph));
66        frame.setVisible(true);
67    }
```

Listing 8.6: A method for displaying the dendrogram

The result is shown in *Figure 8.13*:

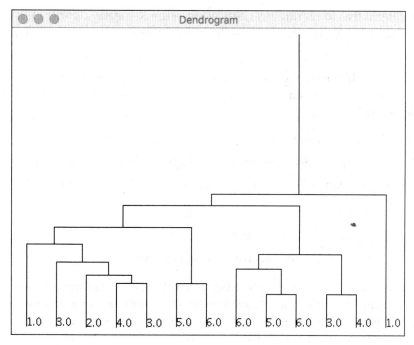

Figure 8.13: Dendrite generated programmatically

This is topologically the same as the tree in *Figure 8.12*.

The key code here is at line 65. The `HierarchyVisualizer()` constructor creates an object from the `graph` string that can be displayed by adding it to the frame's `ContentPane` object this way.

K-means clustering

A popular alternative to hierarchical clustering is the K-means algorithm. It is related to the **K-Nearest Neighbor (KNN)** classification algorithm that we saw in *Chapter 7, Classification Analysis*.

As with hierarchical clustering, the K-means clustering algorithm requires the number of clusters, k, as input. (This version is also called the **K-Means++** algorithm) Here is the algorithm:

1. Select k points from the dataset.
2. Create k clusters, each with one of the initial points as its centroid.
3. For each dataset point x that is not already a centroid:
    - Find the centroid y that is closest to x
    - Add x to that centroid's cluster
    - Re-compute the centroid for that cluster

It also requires k points, one for each cluster, to initialize the algorithm. These initial points can be selected at random, or by some a priori method. One approach is to run hierarchical clustering on a small sample taken from the given dataset and then pick the centroids of those resulting clusters.

This algorithm is implemented in *Listing 8.7*:

```java
public class KMeans {
    private static final double[][] DATA = {{1,1}, {1,3}, {1,5}, {2,6}, {3,2},
            {3,4}, {4,3}, {5,6}, {6,3}, {6,4}, {7,1}, {7,5}, {7,6}};
    private static final int M = DATA.length;    // number of points
    private static final int K = 3;              // number of clusters
    private static HashSet<Point> points;
    private static HashSet<Cluster> clusters = new HashSet();
    private static Random RANDOM = new Random();

    public static void main(String[] args) {
        points = load(DATA);

        // Select a point p at random:
        int i0 = RANDOM.nextInt(M);
        Point p = new Point(DATA[i0][0], DATA[i0][1]);
        points.remove(p);

        // Create a singleton set containing p:
        HashSet<Point> initSet = new HashSet();
        initSet.add(p);

        // Add K-1 more points to initSet:
        for (int i = 1; i < K; i++) {
            p = farthestFrom(initSet);
            initSet.add(p);
            points.remove(p);
        }

        // Create a cluster for each point in the initSet:
        for (Point point : initSet) {
            Cluster cluster = new Cluster(point);
            clusters.add(cluster);
        }

        // Add each remaining point to its nearest (updated) cluster:
        for (Point point : points) {
            Cluster cluster = closestTo(point);
            cluster.add(point);
            cluster.recomputeCentroid();
        }

        System.out.println(clusters);
    }
```

Listing 8.7: K-means clustering

Its `loadData()` method is shown in the preceding *Listing 8.2*. Its other four methods are shown in *Listing 8.8*:

```java
56      /* Returns the cluster whose centroid is closest to the specified point.
57       */
58      private static Cluster closestTo(Point point) {
59          double minDist = Double.POSITIVE_INFINITY;
60          Cluster c = null;
61          for (Cluster cluster : clusters) {
62              double d = distance2(cluster.getCentroid(), point);
63              if (d < minDist) {
64                  minDist = d;
65                  c = cluster;
66              }
67          }
68          return c;
69      }
70
71      /* Returns the point that is farthest from the specified set.
72       */
73      private static Point farthestFrom(Set<Point> set) {
74          Point p = null;
75          double maxDist = 0.0;
76          for (Point point : points) {
77              if (set.contains(point)) {
78                  continue;
79              }
80              double d = dist(point, set);
81              if (d > maxDist) {
82                  p = point;
83                  maxDist = d;
84              }
85          }
86          return p;
87      }
88
89      /* Returns the distance from p to the nearest point in the set:
90       */
91      public static double dist(Point p, Set<Point> set) {
92          double minDist = Double.POSITIVE_INFINITY;
93          for (Point point : set) {
94              double d = distance2(p, point);
95              minDist = (d < minDist? d: minDist);
96          }
97          return minDist;
98      }
99
100     /* Returns the square of the Euclidean distance between the two points.
101      */
102     public static double distance2(Point p, Point q) {
103         double dx = p.getX() - q.getX();
104         double dy = p.getY() - q.getY();
105         return dx*dx + dy*dy;
106     }
```

Listing 8.8: Methods used in Listing 8.7

Output from a run of the program is shown in *Figure 8.14*.

The program loads the data into the `points` set at line 22. Then it selects a point at random and removes it from that set. At lines 29-31, it creates a new point set named `initSet` and adds that random point to it. Then at lines 33-38, it repeats that process for K-1 more points, each selected as the farthest point from the `initSet`. This completes *step 1* of the algorithm. *Step 2* is implemented at lines 40-44, and *step 3* at lines 46-51.

This implementation of *step 1* begins by selecting a point at random. Consequently, different results are likely from separate runs of the program. Note that this output is quite different from the results that we got from hierarchical clustering in *Figure 8.11* above.

The Apache Commons Math library implements this algorithm with its `KMeansPlusPlusClusterer` class, illustrated in *Listing 8.9*:

```java
import org.apache.commons.math3.ml.clustering.CentroidCluster;
import org.apache.commons.math3.ml.clustering.DoublePoint;
import org.apache.commons.math3.ml.clustering.KMeansPlusPlusClusterer;
import org.apache.commons.math3.ml.distance.EuclideanDistance;

public class KMeansPlusPlus {
    private static final double[][] DATA = {{1,1}, {1,3}, {1,5}, {2,6}, {3,2},
        {3,4}, {4,3}, {5,6}, {6,3}, {6,4}, {7,1}, {7,5}, {7,6}};
    private static final int M = DATA.length;   // number of points
    private static final int K = 3;              // number of clusters
    private static final int MAX = 100;          // maximum number of iterations
    private static final EuclideanDistance ED = new EuclideanDistance();

    public static void main(String[] args) {
        List<DoublePoint> points = load(DATA);
        KMeansPlusPlusClusterer<DoublePoint> clusterer;
        clusterer = new KMeansPlusPlusClusterer(K, MAX, ED);
        List<CentroidCluster<DoublePoint>> clusters = clusterer.cluster(points);

        for (CentroidCluster<DoublePoint> cluster : clusters) {
            System.out.println(cluster.getPoints());
        }
    }

    private static List<DoublePoint> load(double[][] data) {
        List<DoublePoint> points = new ArrayList();
        for (int i = 0; i < data.length; i++) {
            points.add(new DoublePoint(data[i]));
        }
        return points;
    }
}
```

```
run:
[[1.0, 1.0], [1.0, 3.0], [1.0, 5.0], [2.0, 6.0], [3.0, 2.0], [3.0, 4.0], [4.0, 3.0]]
[[5.0, 6.0], [6.0, 3.0], [6.0, 4.0], [7.0, 5.0], [7.0, 6.0]]
[[7.0, 1.0]]
```

Listing 8.9: Apache Commons Math K-means++

Cluster Analysis

The output is similar to what we got with our other clustering programs.

A different, more deterministic implementation of *step 1* would be to apply hierarchical clustering first and then select the point in each cluster that is closest to its centroid. For our dataset, we can see from *Figure 8.11*, that would give us the initial set {(3,2), (7,1), (6,4)} or {(3,2), (7,1), (7,5)}, since (6,4) and (7,5) tie for being closest to the centroid (6.2,4.8).

The simplest version of K-means picks all the initial *k* clusters at random. This runs faster than the other two methods described here, but the results are usually not as satisfactory. Weka implements this version with its `SimpleKMeans` class, illustrated in *Listing 8.10*:

```java
import weka.clusterers.SimpleKMeans;
import weka.core.Attribute;
import weka.core.Instance;
import weka.core.Instances;
import weka.core.SparseInstance;

public class KMeans {
    private static final double[][] DATA = {{1,1}, {1,3}, {1,5}, {2,6}, {3,2},
            {3,4}, {4,3}, {5,6}, {6,3}, {6,4}, {7,1}, {7,5}, {7,6}};
    private static final int M = DATA.length;  // number of points
    private static final int K = 3;             // number of clusters

    public static void main(String[] args) {
        Instances dataset = load(DATA);
        SimpleKMeans skm = new SimpleKMeans();
        System.out.printf("%d clusters:%n", K);
        try {
            skm.setNumClusters(K);
            skm.buildClusterer(dataset);
            for (Instance instance : dataset) {
                System.out.printf("(%.0f,%.0f): %s%n",
                        instance.value(0), instance.value(1),
                        skm.clusterInstance(instance));
            }
        } catch (Exception e) {
            System.err.println(e);
        }
    }

    private static Instances load(double[][] data) {
        ArrayList<Attribute> attributes = new ArrayList<Attribute>();
        attributes.add(new Attribute("X"));
        attributes.add(new Attribute("Y"));
        Instances dataset = new Instances("Dataset", attributes, M);
        for (double[] datum : data) {
            Instance instance = new SparseInstance(2);
            instance.setValue(0, datum[0]);
            instance.setValue(1, datum[1]);
            dataset.add(instance);
        }
        return dataset;
    }
}
```

Listing 8.10: K-means with Weka

The output is shown in *Figure 8-15*.

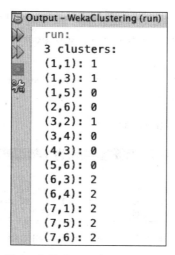

Figure 8.15: Output from Listing 8.9

This program is very similar to that in *Listing 8.5*, where we applied the `HierarchicalClusterer` from the same `weka.clusterers` package. But the results seem less satisfactory. It clusters (7,1) with the four points above it, and it clusters (3,2) with (1,1) and (1,3), but not with (3,4), which is closer.

K-medoids clustering

The k-medoids clustering algorithm is similar to the k-means algorithm, except that each cluster center, called its medoid, is one of the data points instead of being the mean of its points. The idea is to minimize the average distances from the medoids to points in their clusters. The Manhattan metric is usually used for these distances. Since those averages will be minimal if and only if the distances are, the algorithm is reduced to minimizing the sum of all distances from the points to their medoids. This sum is called the **cost of the configuration**.

Here is the algorithm:

1. Select k points from the dataset to be medoids.
2. Assign each data point to its closest medoid. This defines the k clusters.

Cluster Analysis

3. For each cluster C_j:
   - Compute the sum $s = \sum_j s_j$, where each $s_j = \Sigma\{d(\mathbf{x}, \mathbf{y}_j) : \mathbf{x} \in C_j\}$, and change the medoid y_j to whatever point in the cluster C_j that minimizes s
   - If the medoid y_j was changed, re-assign each x to the cluster whose medoid is closest

4. Repeat *step 3* until s is minimal.

```
Output – Clustering (run)
run:
[
{(6.33,4.17),[(7.00,6.00), (7.00,5.00), (5.00,6.00), (6.00,3.00), (6.00,4.00), (7.00,1.00)]},
{(2.25,2.25),[(1.00,1.00), (1.00,3.00), (3.00,2.00), (4.00,3.00)]},
{(2.00,5.00),[(2.00,6.00), (1.00,5.00), (3.00,4.00)]}]
```

Figure 8-14. Output from Listing 8-7

This is illustrated by the simple example in *Figure 8.16*. It shows 10 data points in 2 clusters. The two medoids are shown as filled points. In the initial configuration it is:

$$C_1 = \{(1,1),(2,1),(3,2),(4,2),(2,3)\}, \text{ with } \mathbf{y}_1 = \mathbf{x}_1 = (1,1)$$
$$C_2 = \{(4,3),(5,3),(2,4),(4,4),(3,5)\}, \text{ with } \mathbf{y}_2 = \mathbf{x}_{10} = (3,5)$$

The sums are:

$$s_1 = d(\mathbf{x}_2,\mathbf{y}_1) + d(\mathbf{x}_3,\mathbf{y}_1) + d(\mathbf{x}_4,\mathbf{y}_1) + d(\mathbf{x}_5,\mathbf{y}_1) = 1+3+4+3 = 11$$
$$s_2 = d(\mathbf{x}_6,\mathbf{y}_1) + d(\mathbf{x}_7,\mathbf{y}_1) + d(\mathbf{x}_8,\mathbf{y}_1) + d(\mathbf{x}_9,\mathbf{y}_1) = 3+4+2+2 = 11$$
$$s = s_1 + s_2 = 11+11 = 22$$

The algorithm at *step 3* first part changes the medoid for C_1 to $y_1 = x_3 = (3,2)$. This causes the clusters to change, at *step 3* second part, to:

$$C_1 = \{(1,1),(2,1),(3,2),(4,2),(2,3),(4,3),(5,3)\}, \text{ with } \mathbf{y}_1 = \mathbf{x}_3 = (3,2)$$
$$C_2 = \{(2,4),(4,4),(3,5)\}, \text{ with } \mathbf{y}_2 = \mathbf{x}_{10} = (3,5)$$

This makes the sums:

$$s_1 = 3+2+1+2+2+3 = 13$$
$$s_2 = 2+2 = 4$$
$$s = s_1 + s_2 = 13+4 = 17$$

The resulting configuration is shown in the second panel of *Figure 8-16*:

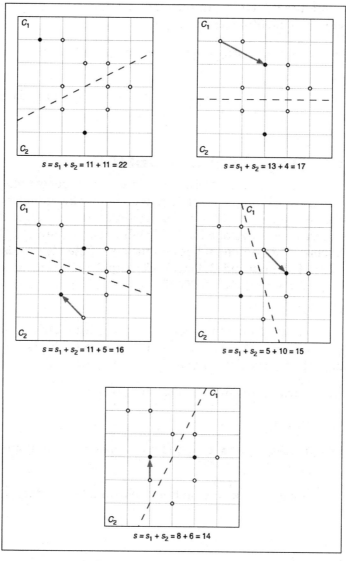

Figure 8.16: K-medoid clustering

At *step 3* of the algorithm, the process repeats for cluster C_2. The resulting configuration is shown in the third panel of *Figure 8.16*. The computations are:

$$C_1 = \{(1,1),(2,1),(3,2),(4,2),(4,3),(5,3)\}, \text{with } \mathbf{y}_1 = \mathbf{x}_3 = (3,2)$$
$$C_2 = \{(2,3),(2,4),(4,4)(3,5)\}, \text{with } \mathbf{y}_2 = \mathbf{x}_8 = (2,4)$$
$$s = s_1 + s_2 = (3+2+1+2+3)+(1+2+2) = 11+5 = 16$$

The algorithm continues with two more changes, finally converging to the minimal configuration shown in the fifth panel of *Figure 8.16*.

This version of k-medoid clustering is also called **partitioning around medoids (PAM)**.

Like K-means clustering, k-medoid clustering is ill-suited for large datasets. It does, however, overcome the problem of outliers, evident in *Figure 8.14*.

Affinity propagation clustering

One disadvantage of each of the clustering algorithms previously presented (hierarchical, k-means, k-medoids) is the requirement that the number of clusters k be determined in advance. The affinity propagation clustering algorithm does not have that requirement. Developed in 2007 by Brendan J. Frey and Delbert Dueck at the University of Toronto, it has become one of the most widely-used clustering methods. (B. J. Frey and D. Dueck, *Clustering by Passing Messages Between Data Points* Science 315, Feb 16, 2007 `http://science.sciencemag.org/content/315/5814/972`).

Like k-medoid clustering, affinity propagation selects cluster center points, called **exemplars**, from the dataset to represent the clusters. This is done by **message-passing** between the data points.

The algorithm works with three two-dimensional arrays:

s_{ij} = the similarity between $\mathbf{x}_i$ and $\mathbf{x}_j$

r_{ik} = responsibility: message from $\mathbf{x}_i$ to $\mathbf{x}_k$ on how well-suited $\mathbf{x}_k$ is as an exemplar for $\mathbf{x}_i$

a_{ik} = availability: message from $\mathbf{x}_k$ to $\mathbf{x}_i$ on how well-suited $\mathbf{x}_k$ is as an exemplar for $\mathbf{x}_i$

We think of the r_{ik} as messages from x_i to x_k, and the a_{ik} as messages from x_k to x_i. By repeatedly re-computing these, the algorithm maximizes the total similarity between the data points and their exemplars.

Figure 8.17 shows how the message-passing works. Data point x_i sends the message r_{ik} to data point x_k, by updating the value of the array element r[i][k]. That value represents how well-suited (from the view of x_i) the candidate x_k would be as an exemplar (representative) of x_i. Later, x_k sends the message a_{ik} to data point xi, by updating the value of the array element a[i][k]. That value represents how well-suited (from the view of x_k) the candidate x_k would be as an exemplar (representative) of x_i. In both cases, the higher the value of the array element, the higher the level of suitability.

The algorithm begins by setting the similarity values to be $s_{ij} = -d(x_i, x_j)^2$, for $i \neq j$, where $d()$ is the Euclidean metric. Squaring the distance simply eliminates the unnecessary step of computing square roots. Changing the sign ensures that $s_{ij} > s_{ik}$ when x_i is closer to x_j than to x_k; i.e., x_i is more similar to x_j than to x_k. For example, in *Figure 8.17*, $x_1 = (2,4)$, $x_2 = (4,1)$, and $x_3 = (5,3)$. Clearly, x_2 is closer to x_3 than to x_1 and $s_{23} > s_{21}$, because $s_{23} = -5 > -13 = s_{21}$.

We also set each s_{ii} to the average of the s_{ij}, for which $i \neq j$. To reduce the number of clusters, that common value can be reduced to the minimum instead of the average of the others. The algorithm then repeatedly updates all the responsibilities r_{ik} and the availabilities a_{ik}.

In general, the suitability of a candidate x_k being an exemplar of a point x_i will be determined by the sum:

$$a_{ik} + r_{ik}$$

This sum measures the suitability of such representation from the view of x_k (availability) combined with that from the view of x_i (responsibility). When that sum converges to a maximum value, it will have determined that representation.

Cluster Analysis

Conversely, the higher $a_{ij}+r_{ij}$ is for some other $j \neq k$, the less suited x_k is as an exemplar of a point x_i. This leads to our update formula for r_{ik}:

$$r_{ik} = s_{ik} - \max\{a_{ij} + s_{ij} : j \neq k\}$$

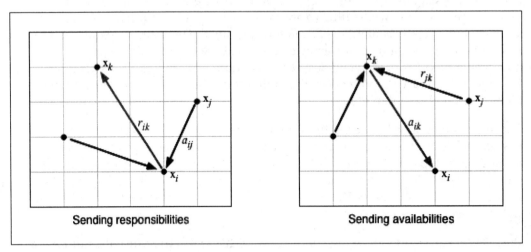

Figure 8.17: Affinity Propagation

For x_k to represent a data point x_i, we want the two points to be similar (high s_{ik}), but we don't want any other x_j to be a better representative (low $a_{ij} + s_{ij}$ for $j \neq k$).

Note that, initially, all the a_{ij} (and the r_{ij}) will be zero. So, on the first iteration:

$$r_{ik} = s_{ik} - \max\{s_{ij} : j \neq k\}$$

That is, each responsibility value is set equal to the corresponding similarity value minus the similarity value of the closest competitor.

Each candidate exemplar x_k measures its availability a_{ik} to represent another data point x_i by adding to its own self-responsibility r_{kk} the sum of the positive responsibilities r_{jk} that it receives from the other points:

$$a_{ik} = \min\{0, r_{kk} + \Sigma\{\max\{0, r_{jk}\} : j \neq i \wedge j \neq k\}\}$$

Note that the sum is thresholded by zero, so that only non-positive values will be assigned to a_{ik}.

The self-availability a_{kk} measuring the confidence that x_k has in representing itself is updated separately:

$$a_{kk} = \Sigma\{\max\{0, r_{jk}\} : j \neq k\}$$

This simply reflects that that self-confidence is accumulated positive confidence (responsibilities) that the other points have for x_k.

Here is the complete algorithm:

1. Initialize the similarities:
   - $s_{ij} = -d(x_i, x_j)^2$, for $i \neq j$;
   - $s_{ii} = $ the average of those other s_{ij} values
2. Repeat until convergence:
   - Update the responsibilities:

   $$r_{ik} = s_{ik} - \max\{a_{ij} + s_{ij} : j \neq k\}$$

   - Update the availabilities:

   $$a_{ik} = \min\{0, r_{kk} + \Sigma_j\{\max\{0, r_{jk}\} : j \neq i \wedge j \neq k\}\}, \text{for } i \neq k;$$
   $$a_{kk} = \Sigma_j\{\max\{0, r_{jk}\} : j \neq k\}$$

A point x_k will be an exemplar for a point x_i if $a_{ik} + r_{ik} = \max_j\{a_{ij} + r_{ij}\}$.

Cluster Analysis

The algorithm is implemented in the program shown in *Listing 8-11*.

```java
public class AffinityPropagation {
    private static double[][] x = {{1,2}, {2,3}, {4,1}, {4,4}, {5,3}};
    private static int n = x.length;                    // number of points
    private static double[][] s = new double[n][n];     // similarities
    private static double[][] r = new double[n][n];     // responsibilities
    private static double[][] a = new double[n][n];     // availabilities
    private static final int ITERATIONS = 10;
    private static final double DAMPER = 0.5;

    public static void main(String[] args) {
        initSimilarities();
        for (int i = 0; i < ITERATIONS; i++) {
            updateResponsibilities();
            updateAvailabilities();
        }
        printResults();
    }

    private static void initSimilarities() {...12 lines}

    private static void updateResponsibilities() {...15 lines}

    private static void updateAvailabilities() {...12 lines}

    /* Returns the negative square of the Euclidean distance from x to y.
     */
    private static double negSqEuclidDist(double[] x, double[] y) {...5 lines}

    /* Returns the sum of the positive r[j][k] excluding r[i][k] and r[k][k].
     */
    private static double sumOfPos(int i, int k) {...9 lines}

    private static void printResults() {...14 lines}
}
```

```
run:
point 0 has exemplar point 1
point 1 has exemplar point 1
point 2 has exemplar point 4
point 3 has exemplar point 4
point 4 has exemplar point 4
```

Listing 8.11: Affinity propagation clustering

It is run on the small dataset {(1,2), (2,3), (4,1), (4,4), (5,3)}, shown in *Figure 8.18*.

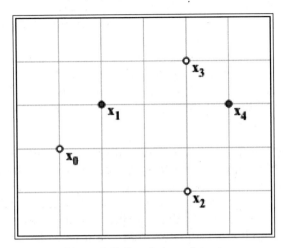

Figure 8.18: Sample input dataset

As the output shows, the five points are organized into two clusters, with exemplars $x_1 = (2,3)$ and $x_4 = (5,3)$.

The `main()` method initializes the similarities array `s[][]` at line 18. Then the main loop repeatedly updates the responsibilities array `r[][]` and the availabilities array `a[][]` at lines 19-22. Finally, the results are printed at line 23.

The `initSimilarities()` method is shown in *Listing 8.12*.

```
26      private static void initSimilarities() {
27          double sum = 0;
28          for (int i = 0; i < n; i++) {
29              for (int j = 0; j < i; j++) {
30                  sum += s[i][j] = s[j][i] = negSqEuclidDist(x[i], x[j]);
31              }
32          }
33          double average = 2*sum/(n*n - n);   // average of s[i][j] for j < i
34          for (int i = 0; i < n; i++) {
35              s[i][i] = average;
36          }
37      }
```

Listing 8.12. Initializing the similarities array

Cluster Analysis

It implements *step 1* of the algorithm. At line 30, the negation of the square of the Euclidean distance between the two points x_i and x_j is assigned to both `s[i][j]` and `s[j][i]`. That value is computed by the auxiliary method `negSqEuclidDist()`, which is defined at lines 68-74 (*Listing 8.16*). The sum of those values is accumulated in the variable `sum`, which is then used at line 33 to compute their mean `average`. (In their original 2007 paper, Frey and Dueck recommend using the median average here. In our implementation, we use the mean average instead). That average value is then re-assigned to all the diagonal elements `s[i][i]` at line 35, as directed by *step 1* of the algorithm.

The initial value that is assigned to the diagonal elements s_{ii} at line 35 can be adjusted to affect the number of exemplars (clusters) that are generated. In their paper, Frey and Dueck show that, with their sample dataset of 25 points, they can obtain a range of results, from one cluster to 25 clusters, by varying that initial value from –100 to –0.1. So, a common practice is to run the algorithm using the average value on that diagonal and then re-run it after adjusting that initial value to generate a different number of clusters.

Note that the assignment to `sum` at line 30 executes $(n^2 - n)/2$ times as `i` iterates from 0 to $n - 1$ and `j` iterates from 0 to $i - 1$ (for example, if $n = 5$, then there will be *20* assignments to `sum`). So, at line 33, the `average` is assigned the `sum` divided by $(n^2 - n)/2$. This is the average of all the elements that lie below the diagonal. Then that constant is assigned to each diagonal element at line 35.

Note that, because of the double assignment at line 30, the array `s[][]` is (as a matrix) symmetric about its diagonal. So, the constant, `average`, is also the average of all the elements above the diagonal.

The `updateResponsibilities()` method is shown in *Listing 8.13*:

```
AffinityPropagation.java
39      private static void updateResponsibilities() {
40          for (int i = 0; i < n; i++) {
41              for (int k = 0; k < n; k++) {
42                  double oldValue = r[i][k];
43                  double max = Double.NEGATIVE_INFINITY;
44                  for (int j = 0; j < n; j++) {
45                      if (j != k) {
46                          max = Math.max(max, a[i][j] + s[i][j]);
47                      }
48                  }
49                  double newValue = s[i][k] - max;
50                  r[i][k] = DAMPER*oldValue + (1 - DAMPER)*newValue;
51              }
52          }
53      }
```

Listing 8.13: Updating the Responsibilities Array

It implements *step 2* first part of the algorithm. At lines 43-48, the value of $max\{a_{ij}+s_{ij}:j\neq k\}$ is computed. That max value is then used to compute $s_{ik} - max\{a_{ij}+s_{ij}:j\neq k\}$ at line 49. That damped value is then assigned to r_{ik} at line 50.

```java
AffinityPropagation.java
55      private static void updateAvailabilities() {
56          for (int i = 0; i < n; i++) {
57              for (int k = 0; k < n; k++) {
58                  double oldValue = a[i][k];
59                  double newValue = Math.min(0, r[k][k] + sumOfPos(i,k));
60                  if (k == i) {
61                      newValue = sumOfPos(k,k);
62                  }
63                  a[i][k] = DAMPER*oldValue + (1 - DAMPER)*newValue;
64              }
65          }
66      }
```

Listing 8-14. Updating the Availabilities Array

The damping performed at line 50 and again at line 63 is recommended by Frey and Dueck to avoid numerical oscillations. They recommend a damping factor of 0.5, which is the value to which the DAMPER constant is initialized at line 15 in *Listing 8.15*.

```java
AffinityPropagation.java
88      private static void printResults() {
89          for (int i = 0; i < n; i++) {
90              double max = a[i][0] + r[i][0];
91              int k = 0;
92              for (int j = 1; j < n; j++) {
93                  double arij = a[i][j] + r[i][j];
94                  if (arij > max) {
95                      max = arij;
96                      k = j;
97                  }
98              }
99              System.out.printf("point %d has exemplar point %d%n", i, k);
100         }
101     }
```

Listing 8.15: Printing the Results

Cluster Analysis

The `updateAvailabilities()` method is shown in *Listing 8.14*. It implements *step 2* second part of the algorithm. The value of $\min\left\{0, r_{kk} + \Sigma\left\{\max\left\{0, r_{jk}\right\} : j \neq i \wedge j \neq k\right\}\right\}$ is computed at line 59. The sum in that expression is computed separately by the auxiliary method `sumOfPos()`, which is defined at lines 76-86 (*Listing 8.16*).

```java
AffinityPropagation.java
68      /* Returns the negative square of the Euclidean distance from x to y.
69      */
70      private static double negSqEuclidDist(double[] x, double[] y) {
71          double d0 = x[0] - y[0];
72          double d1 = x[1] - y[1];
73          return -(d0*d0 + d1*d1);
74      }
75
76      /* Returns the sum of the positive r[j][k] excluding r[i][k] and r[k][k].
77      */
78      private static double sumOfPos(int i, int k) {
79          double sum = 0;
80          for (int j = 0; j < n; j++) {
81              if (j != i && j != k) {
82                  sum += Math.max(0, r[j][k]);
83              }
84          }
85          return sum;
86      }
```

Listing 8.16: Auxiliary Methods

It is the sum of all the positive r_{jk}, excluding r_{ik} and r_{kk}. The element a_{ik} is then assigned that (damped) value at line 63. The diagonal elements a_{kk} are re-assigned the value of `sumOfPos(k,k)` at line 61, as required by *step 2* second part of the algorithm.

The `printResults()` method is shown in *Listing 8.15*. It computes and prints the exemplar (cluster representative) for each point of the dataset. These are determined by the criterion specified in the algorithm: the point x_k is the exemplar for the point x_i if $a_{ik} + r_{ik} = max_j \{a_{ij} + r_{ij}\}$. That index k is computed for each i at lines 90-98 and then printed at line 99.

In their original 2007 paper, Frey and Dueck recommend iterating until the exemplar assignments remain unchanged for 10 iterations. In this implementation, with such a small dataset, we made a total of only 10 iterations.

In his 2009 Ph.D. thesis, D. Dueck mentions that "one run of *k*-medoids may be needed to resolve contradictory solutions." (*Affinity Propagation: Clustering Data by Passing Messages*, U. of Toronto, 2009.)

Summary

After discussing various distance measures and the problems with high-dimensional clustering, this chapter presented four fundamental clustering algorithms: hierarchical clustering, k-means clustering, k-medoids clustering, and the affinity propagation algorithm. These were implemented here in Java, including methods provided by the Weka and the Apache Commons Math libraries.

9
Recommender Systems

Most online shoppers are probably familiar with Amazon's recommender system:

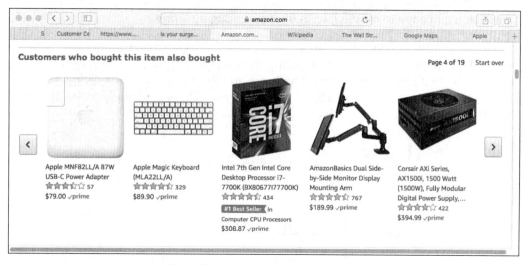

Figure 9.1: Amazon.com recommendations

When a customer views one item, the website displays a list of similar items that have sold well. That comes from their recommender system accessing Amazon's (amazing) database of products, customers, and sales.

Online recommender systems are now run by many vendors of goods and services: Netflix recommending movies, Apple recommending music, Audible recommending books, Yelp recommending restaurants, and so on.

A **recommender system** is an algorithm that predicts a customer's preferences for products based upon an analysis of that customer's previous choices compared to those of many other customers. These algorithms were pioneered by Amazon and Netflix, and are now widely used on the web.

Recommender Systems

Clustering algorithms provide one mechanism for building a recommender system: recommend what the other data points in the same cluster do. More specifically, we could use K-means clustering, and then recommend what the mean of that cluster does. Both clustering algorithms and classification algorithms are used like that to implement recommender systems. In this chapter, however, we will examine several algorithms that are specifically designed for recommender systems.

Utility matrices

Most recommender systems use input that quantify users' preferences for items. These preferences are typically arranged in a matrix that has one row for each user and one column for each item. Such a matrix is called a **utility matrix**. For example, Netflix asks its users to rate movies from one to five stars. So, each entry in that utility matrix would be an integer u_{ij} in the range 0 to 5, representing the number of stars that user i gave to movie j, with 0 representing *no rating*.

For example, *Table 9.1* shows a utility matrix that represents users' ratings of beers on a scale of *1-5*, with 5 representing the greatest approval. Blanks represent no rating by that user for that item. The beers are: BL = *Bud Light*, G = *Guinness*, H = *Heineken*, PU = *Pilsner Urquell*, SA = *Stella Artois*, SNPA = *Sierra Nevada Pale Ale*, and W = *Warsteiner*.

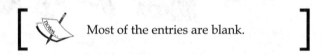

Most of the entries are blank.

	BL	G	H	PU	SA	SNPA	W
x_1		5		4			2
x_2	2		3			5	3
x_3	1		4		3		
x_4	3	4		5		4	
x_5			4		3	3	

Table 9.1: Utility matrix of beer ratings on a 1-5 scale

The purpose of a recommender system is to fill in some of the blanks in the utility matrix. For example, how much would user x_1 like *BL*? To make recommendations, we obviously want to find the blanks whose missing ratings would be high for that user.

Our system might conclude that x_1 would rate BL the same as user x_4 did (three stars) because the two users rated G and PU nearly the same (four stars and five stars). Or, with a lot more data, it might find that most users rate G and $SNPA$ the same, as user x_4 did (four stars), and thus predict that user x_1 would rate them the same, thus giving $SNPA$ five stars.

One approach would be to compare how users rank the properties of the items. For the beer example, we might ask users' preferences for light/dark, bitter/mild, high/low ABV (alcohol by volume), malty, wheaty, and so on. Then we could use those preferences to measure item similarity for each user. This is called **content-based recommendation**.

Alternatively, we could measure the similarity of items by comparing how the users rank the them. This is called **item-based recommendation**. It is usually implemented by means of a **similarity measure**, which is a function $s(y, z)$ that assigns a number to each pair of items (y, z) subject to these properties:

1. $0 \le s(y, z) \le 1$, for all vectors y and z.
2. $s(y, z) = s(z, y)$, for all vectors y and z.
3. $s(y, z) = 1$ when $y = z$.
4. $s(y, z) \approx 1$ when y and z are very similar.
5. $s(y, z) \approx 0$ when y and z are very different.
6. $s(y, z) > s(y, w)$, when y is more similar to z than to w.

If we number the m users $1, 2, \ldots, m$ and the n items $1, 2, \ldots, n$, then we can encapsulate the data with an $m \times n$ utility matrix (u_{ij}) and the $n \times n$ similarity matrix (s_{jk}), where u_{ij} is the rating that user x_i gave item y_j, and s_{jk} is the similarity value $s_{jk} = s(y_j, y_k)$ for the two items y_j and y_k.

With this notation, the ith row of the utility matrix represents all the ratings of user x_i, the jth column of the utility matrix represents all the ratings of item y_j, and the $(j,k)^{th}$ entry of the similarity matrix represents the similarity of items y_j and y_k.

The utility matrix is constructed from the actual list of user ratings that is provided. The similarity matrix is constructed from the utility matrix and the similarity measure.

Similarity measures

A similarity measure is like an inverse distance function. In fact, if $d(y, z)$ is a distance function on the set of all items, we could use this:

$$s(\mathbf{y},\mathbf{z}) = \frac{1}{1+d(\mathbf{y},\mathbf{z})}$$

as a similarity measure. You can check that this would satisfy the six properties for a similarity measure, enumerated previously.

Without a predefined distance function to use, we instead will want to define the similarity measure in terms of the contents of the given utility matrix. There are several ways to do this.

If the utility matrix is Boolean (that is, every entry u_{ij} is either 1 or 0, indicating whether user i has bought an item), then we could adapt the Hamming metric. In this case, each column of the utility matrix is a Boolean vector, indicating which users have bought that item. The **Hamming distance** between two Boolean vectors is the number of vector slots where they differ. For example, if y = (1, 1, 1, 0, 0, 1, 0, 0) and z = (1, 0, 1, 1, 1, 0, 0, 1), then the Hamming distance between y and z is $d_H(y, z) = 5$, because the two vectors disagree in the second, fourth, fifth, sixth, and eighth positions.

The Hamming distance has the special property that $d_H(y, z) \leq n$, where n is the length of the vectors. Consequently, the preceding formula for the derived distance function doesn't quite satisfy property five. However, this alternative formula for the derived distance does work, as the reader may verify:

$$s_H(\mathbf{y},\mathbf{z}) = \frac{n - d_H(\mathbf{y},\mathbf{z})}{n + d_H(\mathbf{y},\mathbf{z})}$$

This version satisfies all six of the requirements for a similarity measure.

Hamming similarity does not work very well when the utility matrix entries are more general decimal values. The problem in that case is that equality of real numbers is highly unlikely. To see that, run this test program:

```
8   public class HammingTest {
9       public static void main(String[] args) {
10          int count = 0;
11          for (int i = 0; i < 1000000000; i++) {
12              long a = Double.doubleToLongBits(Math.random());
13              long b = Double.doubleToLongBits(Math.random());
14              if (a == b) {
15                  ++count;
16              }
17          }
18          System.out.println(count);
19      }
20  }
```

```
Output - RecommenderSystems (run)
run:
0
BUILD SUCCESSFUL (total time: 49 seconds)
```

Listing 9.1: Testing equality of real numbers

This shows that billions of random Double values, even restricted to the interval 0 to 1 and then converted to type long integer, will not be equal. So, the Hamming distance between random vectors of real numbers will almost always be n, the length of the vectors.

Cosine similarity

If we think of each column **y** of the utility matrix as an n-dimensional vector, **y** = $(y_1, y_2, ..., y_n)$, then we can use the Euclidean dot product (inner product) formula to compute the cosine of the angle θ that the two vectors make at the origin:

$$\mathbf{y} \cdot \mathbf{z} = |\mathbf{y}||\mathbf{z}|\cos\theta$$

$$\cos\theta = \frac{\mathbf{y} \cdot \mathbf{z}}{|\mathbf{y}||\mathbf{z}|} = \frac{\sum_{j=1}^{n} y_j z_j}{\sqrt{\sum_{j=1}^{n} y_j^2}\sqrt{\sum_{j=1}^{n} z_j^2}}$$

This is called the **cosine similarity** measure:

For example, if $y = (2, 1, 3)$ and $z = (1, 3, 2)$, then:

$$s(\mathbf{y},\mathbf{z}) = \cos\theta = \frac{2\cdot 1 + 1\cdot 3 + 3\cdot 2}{\sqrt{2^2+1^2+3^2}\sqrt{1^2+3^2+2^2}} = \frac{11}{14} = 0.7857$$

We can see that the cosine similarity measure has the six requisite properties for a similarity measure. If u and v are parallel, then $s(y, z) = \cos \theta = \cos 0 = 1$. That would be the result in the case where $y = (2, 1, 2)$ and $z = (4, 2, 4)$. On the other hand, if $y = (2, 0, 2)$ and $z = (0, 4, 0)$, then y and z are perpendicular and $s(y, z) = \cos \theta = \cos 90° = 0$.

We can interpret these extremes in terms of a utility matrix. If $y = (2, 1, 2)$ and $z = (4, 2, 4)$, then $z = 2y$. They are very similar in that all three of the users rated item z twice as high as item y. But in the second example of $(2, 0, 2)$ and $(0, 4, 0)$, we can detect no similarity at all in this data: item y is rated only by users who didn't rate item z, and vice versa.

That example of $u = (2, 0, 2)$ and $v = (0, 4, 0)$ is not as easy to interpret because the value 0 means no evaluation; that is, the user has no opinion on that item. A better example would be when $u = (4, 1, 1)$ and $v = (1, 4, 4)$. Here, u likes item 1 very much and items 2 and 3 hardly at all, while u has just the opposite opinions. Accordingly, their cosine similarity is low: $s(u, v) = 4/11 = 0.3636$.

A simple recommender system

Here is an item-to-item version in which the utility matrix is Boolean:

$u_{ij} = 1 \Leftrightarrow$ user i bought item j.

Recommender algorithm 1 is as follows. Given an input list of (i, j) pairs, representing purchases of items y_j bought by users x_i:

1. Initialize the utility matrix (u_{ij}) with m rows and n columns, where m is the number of users and n is the number of items.
2. For each pair (i, j) in the input list, set $u_{ij} = 1$.
3. Initialize the similarity matrix (s_{jk}) with n rows and n columns.
4. For each $j = 1\ldots n$ and each $k = 1\ldots n$, set $s_{jk} = s(u, v)$, the cosine similarity of the jth column u and the kth column v of the utility matrix.
5. For a given user-purchase pair (i, j) (that is, $u_{ij} = 1$):
    ○ Find the set S of items not bought by user i
    ○ Sort the items in S according to how similar they are to item j
6. Recommend the top n_1 elements of S, where n_1 is a specified constant much less than n.

To implement this algorithm, we need an input list of purchases. The program in *Listing 9.2* generates a random list of pairs, where the pair *(i, j)* represents the purchase of item *j* by user *i*:

```java
public class DataGenerator1 {
    static final Random RANDOM = new Random();
    static final int NUM_USERS = 5;
    static final int NUM_ITEMS = 12;
    static final int NUM_PURCHASES = 36;

    public static void main(String[] args) {
        HashSet<Purchase> purchases = new HashSet(NUM_PURCHASES);
        while (purchases.size() < NUM_PURCHASES) {
            purchases.add(new Purchase());
        }

        File outFile = new File("data/Purchases.dat");
        try {
            PrintWriter out = new PrintWriter(outFile);
            out.printf("%d users%n", NUM_USERS);
            out.printf("%d items%n", NUM_ITEMS);
            out.printf("%d purchases%n", NUM_PURCHASES);
            for (Purchase purchase : purchases) {
                out.println(purchase);
                System.out.println(purchase);
            }
            out.close();
        } catch (FileNotFoundException e) {
            System.err.println(e);
        }
    }

    static class Purchase {...32 lines }
}
```

Listing 9.2: Program to generate user-item pairs

Recommender Systems

The program generates 36 *(i, j)* pairs, where $1 \le i \le 5$ and $1 \le j \le 12$, representing 36 purchases with `5` users and `12` items. These are stored in the external file `Purchases1.dat`, shown in *Figure 9.2*:

```
Purchases1.dat
 1  5 users
 2  12 items
 3  36 purchases
 4   5   1
 5   3   6
 6   1  11
 7   3   7
 8   4   5
 9   2  10
10   5   3
11   5   4
12   3   9
13   4   7
14   2  12
15   5   6
16   3  11
17   1   3
18   4   9
19   2   1
20   5   7
21   1   4
22   3  12
23   2   2
24   1   5
25   5   9
26   3   1
27   2   4
28   5  10
29   3   2
30   1   7
31   2   5
32   5  11
33   3   3
34   1   8
35   2   6
36   5  12
37   2   7
38   3   5
39   1  10
```

Figure 9.2: Purchases file

It also inserts three header lines specifying the numbers of `users`, `items`, and `purchases`.

The static nested `Purchase` class (lines 42-73) is shown in *Listing 9.3*:

```java
       static class Purchase {
           int user;
           int item;

           public Purchase() {
               this.user = RANDOM.nextInt(NUM_USERS) + 1;
               this.item = RANDOM.nextInt(NUM_ITEMS) + 1;
           }

           @Override
           public int hashCode() {
               return NUM_ITEMS*this.user + NUM_USERS*this.item;
           }

           @Override
           public boolean equals(Object object) {
               if (object == null) {
                   return false;
               } else if (object == this) {
                   return true;
               } else if (!(object instanceof Purchase)) {
                   return false;
               }
               Purchase that = (Purchase)object;
               return that.user == this.user && that.item == this.item;
           }

           @Override
           public String toString() {
               return String.format("%4d%4d", user, item);
           }
       }
```

Listing 9.3: The Purchase class in the DataGenerator1 program

The random values for the `user` and `item` numbers are generated in the default constructor at lines 47 and 48. The `hashCode()` and `equals()` methods are included in case the objects are to be used in a `Set` or `Map` collection.

The `Filter1` program in *Listing 9.4* implements steps 1-4 of the algorithm, filtering the `Purchases1.dat` file to generate the `Utility1.dat` and `Similarity1.dat` files to store the utility and similarity matrices. Its methods for computing and storing the matrices are shown in *Listing 9.5* and *Listing 9.6*. The utility matrix entries are read directly from the `Purchases1.dat` file. The similarity matrix entries are computed by the `cosine()` method, shown with its auxiliary `dot()` and `norm()` methods in *Listing 9.7*. The resulting `Utility1.dat` and `Similarity1.dat` files are shown in *Figure 9.3* and *Figure 9.4*.

Recommender Systems

The code at lines 90-91 in *Listing 9.7* prevents division by zero. If the value of `denominator` is zero, then either the jth column or the kth column of the utility matrix is the zero vector; that is, all its components are zero. In that case, the cosine similarity will be *0.0*, by default.

Steps 5 and 6 of the *recommender algorithm 1* are implemented by the `Recommender1` program in *Listing 9.8*. Its methods are shown separately in *Listing 9.9* through *Listing 9.12*. Two sample runs of the program are shown in *Figure 9.5* and *Figure 9.6*:

```java
    public static int[][] computeUtilityMatrix(File file)
            throws FileNotFoundException {
        Scanner in = new Scanner(file);
        // Read the five header lines:
        m = in.nextInt();  in.nextLine();
        n = in.nextInt();  in.nextLine();
        in.nextLine();  in.nextLine();  in.nextLine();

        // Read in the utility matrix:
        int[][] u = new int[m+1][n+1];
        while (in.hasNext()) {
            int i = in.nextInt();   // user
            int j = in.nextInt();   // item
            u[i][j] = 1;
        }
        in.close();
        return u;
    }

    public static void storeUtilityMatrix(int[][] u, File file)
            throws FileNotFoundException {
        PrintWriter out = new PrintWriter(file);
        out.printf("%d users%n", m);
        out.printf("%d items%n", n);
        for (int i = 1; i <= m; i++) {
            for (int j = 1; j <= n; j++) {
                out.printf("%2d", u[i][j]);
            }
            out.println();
        }
        out.close();
    }
```

Listing 9.4: Program to filter the purchases list

```
 13     public class Filter1 {
 14         private static int m;    // number of users
 15         private static int n;    // number of items
 16
 17         public static void main(String[] args) {
 18             File purchasesFile = new File("data/Purchases1.dat");
 19             File utilityFile = new File("data/Utility1.dat");
 20             File similarityFile = new File("data/Similarity1.dat");
 21             try {
 22                 int[][] u = computeUtilityMatrix(purchasesFile);
 23                 storeUtilityMatrix(u, utilityFile);
 24                 double[][] s = computeSimilarityMatrix(u);
 25                 storeSimilarityMatrix(s, similarityFile);
 26             } catch (FileNotFoundException e) {
 27                 System.err.println(e);
 28             }
 29         }
 30
 31         public static int[][] computeUtilityMatrix(File file)
 32                 throws FileNotFoundException {...17 lines }
 49
 50         public static void storeUtilityMatrix(int[][] u, File file)
 51                 throws FileNotFoundException {...12 lines }
 63
 64         public static double[][] computeSimilarityMatrix(int[][] u) {...9 lines }
 73
 74         public static void storeSimilarityMatrix(double[][] s, File file)
 75                 throws FileNotFoundException {...11 lines }
 86
 87         /* Returns the cosine similarity of the jth and kth columns of u[][].
 88          */
 89         public static double cosine(int[][] u, int j, int k) {...3 lines }
 92
 93         /* Returns the dot product of the jth and kth columns of u[][].
 94          */
 95         public static double dot(int[][] u, int j, int k) {...7 lines }
102
103         /* Returns the norm of the jth column of u[][].
104          */
105         public static double norm(int[][] u, int j) {...3 lines }
108     }
```

Listing 9.5: Compute and store the utility matrix

```java
public static double[][] computeSimilarityMatrix(int[][] u) {
    double[][] s = new double[n+1][n+1];
    for (int j = 1; j <= n; j++) {
        for (int k = 1; k <= n; k++) {
            s[j][k] = cosine(u, j, k);
        }
    }
    return s;
}

public static void storeSimilarityMatrix(double[][] s, File file)
        throws FileNotFoundException {
    PrintWriter out = new PrintWriter(file);
    out.printf("%d items%n", n);
    for (int i = 1; i <= n; i++) {
        for (int j = 1; j <= n; j++) {
            out.printf("%6.2f", s[i][j]);
        }
        out.println();
    }
    out.close();
}
```

Listing 9.6: Compute and store the similarity matrix

```java
/* Returns the cosine similarity of the jth and kth columns of u[][].
 */
public static double cosine(int[][] u, int j, int k) {
    double denominator = norm(u,j)*norm(u,k);
    return (denominator == 0 ? 0 : dot(u,j,k)/denominator);
}

/* Returns the dot product of the jth and kth columns of u[][].
 */
public static double dot(int[][] u, int j, int k) {
    double sum = 0.0;
    for (int i = 0; i <= m; i++) {
        sum += u[i][j]*u[i][k];
    }
    return sum;
}

/* Returns the norm of the jth column of u[][].
 */
public static double norm(int[][] u, int j) {
    return Math.sqrt(dot(u,j,j));
}
}
```

Listing 9.7: Methods for computing similarity

Chapter 9

```
    ...java   Utility1.dat  ⊗    Similarity1.
1   5 users
2   12 items
3   0 0 1 1 1 0 1 1 0 1 1 0
4   1 1 0 1 1 1 1 0 0 1 0 1
5   1 1 1 0 1 0 1 0 1 0 1 1
6   0 0 0 0 1 0 1 0 1 0 0 0
7   0 0 1 1 0 1 1 0 1 1 1 1
```

Figure 9.3: The Utility.dat file

```
    ...java   Utility1.dat ⊗   Similarity1.dat ⊗   Recommender1.java ⊗   HammingTest.java ⊗
1   12 items
2    1.00  1.00  0.41  0.41  0.71  0.50  0.63  0.00  0.41  0.41  0.41  0.82
3    1.00  1.00  0.41  0.41  0.71  0.50  0.63  0.00  0.41  0.41  0.41  0.82
4    0.41  0.41  1.00  0.67  0.58  0.41  0.77  0.58  0.67  0.67  1.00  0.67
5    0.41  0.41  0.67  1.00  0.58  0.82  0.77  0.58  0.33  1.00  0.67  0.67
6    0.71  0.71  0.58  0.58  1.00  0.35  0.89  0.50  0.58  0.58  0.58  0.58
7    0.50  0.50  0.41  0.82  0.35  1.00  0.63  0.00  0.41  0.82  0.41  0.82
8    0.63  0.63  0.77  0.77  0.89  0.63  1.00  0.45  0.77  0.77  0.77  0.77
9    0.00  0.00  0.58  0.58  0.50  0.00  0.45  1.00  0.00  0.58  0.58  0.00
10   0.41  0.41  0.67  0.33  0.58  0.41  0.77  0.00  1.00  0.33  0.67  0.67
11   0.41  0.41  0.67  1.00  0.58  0.82  0.77  0.58  0.33  1.00  0.67  0.67
12   0.41  0.41  1.00  0.67  0.58  0.41  0.77  0.58  0.67  0.67  1.00  0.67
13   0.82  0.82  0.67  0.67  0.58  0.82  0.77  0.00  0.67  0.67  0.67  1.00
```

Figure 9.4: The Similarity.dat file

```
  DataGenerator1.java ⊗   Purchases1.dat ⊗   Filter1.java ⊗   Utility1.dat ⊗   Similarity1.dat ⊗   Recommender1.java ⊗
14   public class Recommender1 {
15       private static int m;               // number of users
16       private static int n;               // number of items
17       private static int[][] u;           // utility matrix
18       private static double[][] s;        // similarity matrix
19       private static int user;            // the current user
20       private static int bought;          // the current item bought by user
21
22  ⊞    public static void main(String[] args) {...6 lines }
28
29  ⊞    public static void readFiles() {...10 lines }
39
40  ⊞    public static void readUtilMatrix(File f) throws FileNotFoundException {...13 lines }
53
54  ⊞    public static void readSimilMatrix(File f) throws FileNotFoundException {...13 lines }
67
68  ⊞    public static void getInput() {...9 lines }
77
78  ⊞    private static Set<Item> itemsNotYetBought() {...9 lines }
87
88  ⊞    private static void makeRecommendations(Set<Item> set, int numRecs) {...11 lines }
99
100 ⊞    static class Item implements Comparable<Item> {...19 lines }
119  }
```

Listing 9.8: The Recommender1 program

```
...dat  Filter1.java ×    Utility1.dat ×    Similarity1.dat ×   Recommender1.java ×
22        public static void main(String[] args) {
23            readFiles();
24            getInput();
25            Set<Item> set = itemsNotYetBought();
26            makeRecommendations(set, n/4);
27        }
28
29        public static void readFiles() {
30            File utilityFile = new File("data/Utility1.dat");
31            File similarityFile = new File("data/Similarity1.dat");
32            try {
33                readUtilMatrix(utilityFile);
34                readSimilMatrix(similarityFile);
35            } catch (FileNotFoundException e) {
36                System.err.println(e);
37            }
38        }
```

Listing 9.9: Methods for the Recommender1 program

Step 5 of the algorithm is implemented by the method called at line 25, and *step 6* is implemented by the `makeRecommendations()` method called at line 26. Note that we have chosen the value n/4 for n_1.

The `getInput()` method (*Listing 9.11*) interactively reads a user number and item number, representing a new purchase. Then the `itemsNotYetBought()` method returns a set of `Item` objects that have not yet been bought by that user. Since we are using the `TreeSet<Item>` class, the set remains sorted after each addition (at line 82). The ordering is done in accordance with the overridden `compareTo()` method defined at lines 107-112 in *Listing 9.12*. When two items are compared, the one whose similarity to the bought item is greater will precede the other. This implements *step 5*, the second part of the algorithm. Consequently, when the set is passed to the `makeRecommendation()` method at line 26, its elements are already ordered in decreasing similarity to the bought item.

```java
public static void readUtilMatrix(File f) throws FileNotFoundException {
    Scanner in = new Scanner(f);
    m = in.nextInt();  in.nextLine();
    n = in.nextInt();  in.nextLine();
    u = new int[m+1][n+1];
    for (int i = 1; i <= m; i++) {
        for (int j = 1; j <= n; j++) {
            u[i][j] = in.nextInt();
        }
        in.nextLine();
    }
    in.close();
}

public static void readSimilMatrix(File f) throws FileNotFoundException {
    Scanner in = new Scanner(f);
    n = in.nextInt();
    in.nextLine();
    s = new double[n+1][n+1];
    for (int j = 1; j <= n; j++) {
        for (int k = 1; k <= n; k++) {
            s[j][k] = in.nextDouble();
        }
        in.nextLine();
    }
    in.close();
}
```

Listing 9.10: File reading methods for Recommender1 program

```java
public static void getInput() {
    Scanner input = new Scanner(System.in);
    System.out.print("Enter user number: ");
    user = input.nextInt();
    System.out.print("Enter item number: ");
    bought = input.nextInt();
    System.out.printf("User %d bought item %d.%n", user, bought);
    u[user][bought] = 1;
}

private static Set<Item> itemsNotYetBought() {
    Set<Item> set = new TreeSet();
    for (int j = 1; j <= n; j++) {
        if (u[user][j] == 0) {  // user has not yet bought item j
            set.add(new Item(j));
        }
    }
    return set;
}
```

Listing 9.11: Methods for Recommender1 program

Recommender Systems

```java
 88      private static void makeRecommendations(Set<Item> set, int numRecs) {
 89          System.out.printf("We also recommend these %d items:", numRecs);
 90          int count = 0;
 91          for (Item item : set) {
 92              System.out.printf("  %d", item.index);
 93              if (++count == numRecs) {
 94                  break;
 95              }
 96          }
 97          System.out.println();
 98      }
 99
100      static class Item implements Comparable<Item> {
101          int index;
102
103          public Item(int index) {
104              this.index = index;
105          }
106
107          @Override
108          public int compareTo(Item item) {
109              double s1 = s[bought][this.index];
110              double s2 = s[bought][item.index];
111              return (s1 > s2 ? -1 : 1);
112          }
113
114          @Override
115          public String toString() {
116              return String.format("%d", index);
117          }
118      }
119  }
```

Listing 9.12: Nested Item class for Recommender1 program

The nested `Item` class implements the `Comparable<Item>` interface, which is necessary for using the `Set<Item>` interface. That, in turn, requires the overriding of the `compareTo(Item)` method (at lines 107-112). Lines 109-110 define `s1` and `s2` to be the similarity values (from the similarity matrix) of the implicit argument (`this`) and the explicit argument (`item`) to the `bought` item. If `s1 > s2`, meaning that `this` is more similar than `item` is to the `bought` item, then the method returns `-1`, which means that `this` should precede `item` in the set's ordering. That keeps the elements that are more similar to the `bought` item in front of the others. Consequently, when the for-each loop at lines 91-96 iterates through the first `numRecs` of the set, those elements get recommended.

The first sample run of the `Recommender1` program is shown in *Figure 9.5*:

```
Output - RecommenderSystems (run)
run:
Enter user number: 1
Enter item number: 1
User 1 bought item 1.
We also recommend these 3 items:   2  12   6
```
Figure 9.5: Run 1 of the Recommender1 program

The interactive input tells the program that user #1 has bought item #1. The utility matrix (*Figure 9.3*) shows that, other than *item #1*, *item #2*, *item #6*, *item #9*, and *item #12* have not yet been bought by *user #1*. Thus, the set defined at line 25 is {2, 6, 9, 12}. But the set is ordered according to the similarity matrix (*Figure 9.4*), which shows that the similarities of those four items are *1.00, 0.50, 0.41*, and *0.82* in the first row (corresponding to bought *item #1*). Thus, as an ordered set, it is (2, 12, 6, 9). The value of n/4 is 12/4 = 3, so the `makeRecommendations()` method at line 26 prints the first three items of that ordered set: *2, 12,* and *6*.

The second sample run of the program is shown in *Figure 9.6*:

```
Output - RecommenderSystems (run)
run:
Enter user number: 2
Enter item number: 3
User 2 bought item 3.
We also recommend these 3 items:  11   9   8
```
Figure 9.6: Run 2 of the Recommender1 program

This time, *user #2* has bought *item #3*. The other items that *user #2* has not yet bought are *item #8*, *item #9*, and *item #11*. Their similarities to the bought *item #3* are *0.58, 0.67,* and *1.00*, respectively, so the ordered set is (11, 9, 8). Those three are then recommended.

In the first run, the similarity of *item #2* to *item #1* was found to be 1.00 (that is, *100* percent). That fact can be seen directly from the utility matrix: *column #2* and *column #1* are identical – they're both *(0, 1, 1, 0, 0)*. In the second run, *item #3* and *item #11* have *100* percent similarity: both columns are *(1, 0, 1, 0, 1)*. In both runs, the first recommendation is an item that was uniformly rated the same as the one that the user has just bought.

Amazon's item-to-item collaborative filtering recommender

The recommender algorithm that Amazon adopted early on was an improvement of the `Recommender1` implemented previously. The main difference is in *step 5*, where now we have two more substeps, which pick the n_1 most similar items and sort them by popularity:

Recommender algorithm 2 is as follows:

1. Initialize the utility matrix (u_{ij}) with m rows and n columns, where m is the number of users and n is the number of items.
2. For each pair (i, j) in the input list, set $u_{ij} = 1$.
3. Initialize the similarity matrix (s_{jk}) with n rows and n columns.
4. For each $j = 1...n$ and each $k = 1...n$, set $s_{jk} = s(u, v)$, the cosine similarity of the jth column u and the kth column v of the utility matrix.
5. For a given user-purchase pair (i, j) (that is, $u_{ij} = 1$):
    - Find the set S of items not bought by user i
    - Sort the items in S according to how similar they are to item j
    - Let S' be the top n_1 elements of S
    - Sort the items in S' by popularity
6. Recommend the top n_2 items in S'.

This is implemented in the `Recommender2` program, shown in *Listing 9.13*:

```
15    public class Recommender2 {
16        private static int m;              // number of users
17        private static int n;              // number of items
18        private static int[][] u;          // utility matrix
19        private static double[][] s;       // similarity matrix
20        private static int user;           // the current user
21        private static Item itemBought;    // the current item bought by user
22
23        public static void main(String[] args) {...7 lines }
30
31        public static void readFiles() {...10 lines }
41
42        public static void readUtilMatrix(File f) throws FileNotFoundException {...13 lines }
55
56        public static void readSimilMatrix(File f) throws FileNotFoundException {...13 lines }
69
70        public static void getInput() {...14 lines }
84
85        private static Set<Item> itemsNotYetBought() {...9 lines }
94
95        private static Set<Item> firstPartOf(Set<Item> set1, int n1) {...11 lines }
106
107       private static void makeRecommendations(Set<Item> set, int n2) {...11 lines }
118
119       static class Item {...42 lines }
161   }
```

Listing 9.13: The Recommender2 program

If you compare that with the outline in *Listing 9.8*, you can see the similarities between the two programs. In fact, the three input methods `readFiles()`, `readUtilMatrix()`, and `readSimulMatrix()`, are the same.

Structurally, the main difference between the two programs is in the nested `Item` class. In `Recommender1`, the `Item` class (*Listing 9.12*) implements `Comparable<Item>` with a `compareTo()` method that causes the `TreeSet` (at line 79 in *Listing 9.11*) to be ordered according to similarity to the bought item. But in `Recommender2`, we have to have that ordering to implement the second part of *step 5* and then also reorder the set by item popularity to implement the fourth part of *step 5*. The `Item` class now has to have two different ordering mechanisms: first by similarity and then by popularity.

In Java, a class can have only one `compareTo()` method. When it needs more than one way to compare its elements, it must implement an inner `Comparator` class for each method. This is done in *Listing 9.14*: an inner `PopularityComparator` class at lines 138-145 and an inner `SimilarityComparator` class at lines 147-154, each with its own `compare()` method. These are implemented with the corresponding methods: `popularity()` at lines 126-132 and `similarity()` at lines 134-136. Note that both methods access the field `this.index` (at lines 129 and 135), which is the implicit argument's index in the utility and similarity matrices.

Recommender Systems

In the `popularity()` method, the `for` loop at lines 128-130 computes the number of users who have bought the current item. That `sum` is then used by the corresponding `compare()` method to order the elements of `set` by that measure of popularity.

The `similarity()` method returns the similarity between the current item (the implicit argument) and the specified `item` passed to it. That cosine value is then used by the corresponding `compare()` method the same way that the `compareTo()` method was computed in the `Recommender1` program.

The `getInput()` and `itemsNotYetBought()` methods for the `Recommender2` program, shown in *Listing 9.15*, are nearly the same as those for the `Recommender1` program:

```java
    static class Item {
        int index;

        public Item(int index) {
            this.index = index;
        }

        public int popularity() {
            int sum = 0;
            for (int i = 1; i <= m; i++) {
                sum += u[i][this.index];
            }
            return sum;
        }

        public double similarity(Item item) {
            return s[this.index][item.index];
        }

        public class PopularityComparator implements Comparator<Item> {
            @Override
            public int compare(Item item1, Item item2) {
                int p1 = item1.popularity();
                int p2 = item2.popularity();
                return (p1 > p2 ? -1 : 1);
            }
        }

        public class SimilarityComparator implements Comparator<Item> {
            @Override
            public int compare(Item item1, Item item2) {
                double s1 = Item.this.similarity(item1);
                double s2 = Item.this.similarity(item2);
                return (s1 > s2 ? -1 : 1);
            }
        }

        @Override
        public String toString() {
            return String.format("%d", index);
        }
    }
}
```

Listing 9.14: Item class for Recommender2 program

```java
public static void getInput() {
    Scanner input = new Scanner(System.in);
    System.out.print("Enter user number: ");
    user = input.nextInt();
    System.out.print("Enter item number: ");
    int bought = input.nextInt();
    if (u[user][bought] == 1) {
        System.out.printf("User %d already has item %d.%n", user, bought);
        System.exit(0);
    }
    System.out.printf("User %d bought item %d.%n", user, bought);
    u[user][bought] = 1;
    itemBought = new Item(bought);
}

private static Set<Item> itemsNotYetBought() {
    Set<Item> set = new TreeSet(itemBought.new SimilarityComparator());
    for (int j = 1; j <= n; j++) {
        if (u[user][j] == 0) {  // user has not yet bought item j
            set.add(new Item(j));
        }
    }
    return set;
}
```

Listing 9.15: Methods for the Recommender2 program

The only difference in this `getInput()` method is the addition of lines 76-79 that check whether the specified item had already been bought by the specified user.

In the `itemsNotYetBought()` method, the inner `SimilarityComparator` class is bound to the `itemBought` object at line 86. This is done by means of the rather unusual expression `itemBought.new` preceding the default constructor invocation `SimilarityComparator()`. That is how the `itemBought` object becomes the implicit argument for that class's `compare()` method when bound to the `set` object that is being constructed at line 86. In other words, the `Item.this` reference used at lines 150 and 151 will refer to that `itemBought` object.

Recommender Systems

Listing 9.16 shows the `firstPartOf()` and the `makeRecommendations()` methods:

```java
 95    private static Set<Item> firstPartOf(Set<Item> set1, int n1) {
 96        Set<Item> set2 = new TreeSet(itemBought.new PopularityComparator());
 97        int count = 0;
 98        for (Item item : set1) {
 99            set2.add(item);
100            if (++count == n1) {
101                break;
102            }
103        }
104        return set2;
105    }
106
107    private static void makeRecommendations(Set<Item> set, int n2) {
108        System.out.printf("We also recommend these %d items:", n2);
109        int count = 0;
110        for (Item item : set) {
111            System.out.printf("  %d", item.index);
112            if (++count == n2) {
113                break;
114            }
115        }
116        System.out.println();
117    }
```

Listing 9.16: Other methods for the Recommender2 program

The `itemsNotYetBought()` method returns the set of items that the user has not yet bought, ordered by their similarity to the `itemBought` item. The `firstPartOf()` method picks the first `n1` of those items, returning them in a separate set that is ordered by their popularity.

The `main()` method is shown in *Listing 9.17*:

```java
23    public static void main(String[] args) {
24        readFiles();
25        getInput();
26        Set<Item> set1 = itemsNotYetBought();
27        Set<Item> set2 = firstPartOf(set1, n/3);
28        makeRecommendations(set2, n/4);
29    }
```

Listing 9.17: The main method for the Recommender2 program

Line 26 implements the first two parts of *step 5* of the algorithm, creating `set1` of all items not yet bought by the current `user`. As we have seen, that set is ordered according to similarity with the current item, `itemBought`. Line 27 implements the third and fourth parts of *steps 5*, creating `set2`, the subset of the first n/3 elements of `set1`. And, as we have seen, that set is ordered according to the popularity of those items. Finally, Line 28 implements *step 6*, recommending the first n/4 elements of `set2`.

The choices of n/3 and n/4 for n_1 and n_2 here are rather arbitrary. In general, they should depend in some way upon n, the total number of items. And, of course, we should have $n_2 < n_1 \ll n$. In our test run choices, these values are $n_1 = 4$ and $n_2 = 3$.

Sample runs of the Recommender2 program are shown in *Figure 9.7* and *Figure 9.8*:

```
Output - RecommenderSystems (run)
run:
Enter user number: 1
Enter item number: 1
User 1 bought item 1.
We also recommend these 3 items:  12  9  2
```

Figure 9.7: Run 1 of the Recommender2 program

```
Output - RecommenderSystems (run)
run:
Enter user number: 4
Enter item number: 1
User 4 bought item 1.
We also recommend these 3 items:  12  3  2
```

Figure 9.8: Run 2 of the Recommender2 program

The first run uses the same input that we used for the Recommender1 program (*Figure 9.5*): *user #1* and *item #1*. The result here (12, 9, 2) is different from the previous result of (2, 12, 6). That is because *item #12* and *item #9* are more popular than *item #2* and *item #6*. We can see from the utility matrix (*Figure 9.9*) that *item #12* and *item #9* were bought by three other users, while *item #2* and *item #6* were bought by only two other users.

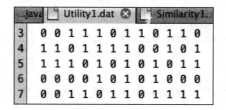

Figure 9.9: Utility matrix

In the second run, *user #4* buys *item #1*. Row #4 of the utility matrix is (0, 0, 0, 0, 1, 0, 1, 0, 1, 0, 0, 0). Other than *column #1*, this shows 0s in columns 2, 3, 4, 6, 8, 10, 11, and 12. Ordered by similarity to *item #1*, those elements form set1 as (2, 12, 6, 3, 4, 10, 11, 8). The algorithm selects the first n_1 (four) of these for set2: {2, 12, 6, 3}. But these are ordered by popularity, so the actual set is set2 = (12, 3, 2, 6). Finally, the algorithm selects the first n_2 (three) of these to be recommended.

Implementing user ratings

Many online vendors ask their customers to rate the products that they purchase, typically on a scale of one to five stars. We can modify our previous `Recommender2` program to incorporate these numerical ratings. To test the new version, we'll also modify our `DataGenerator` and `Filter` programs.

The modified `DataGenerator` program is shown in *Listing 9.18*:

```java
public class DataGenerator3 {
    static final Random RANDOM = new Random();
    static final int NUM_USERS = 5;
    static final int NUM_ITEMS = 12;
    static final int MAX_RATING = 5;
    static final int NUM_PURCHASES = 36;
    static final double MU = 3.0;        // average rating
    static final double SIGMA = 1.0;     // standard deviation

    public static void main(String[] args) {...21 lines}

    static class Purchase {
        int user;
        int item;
        double rating;

        public Purchase() {
            this.user = RANDOM.nextInt(NUM_USERS) + 1;
            this.item = RANDOM.nextInt(NUM_ITEMS) + 1;
            this.rating = randomRating();
        }

        public double randomRating() {
            double x =  MU + SIGMA*RANDOM.nextGaussian();
            x = Math.max(1, x);              // x >= 1.0
            x = Math.min(MAX_RATING, x);     // x <= 5.0
            return Math.floor(2*x)/2;        // 0.5, 1.0, 1.5, 2.0, ..
        }

        @Override
        public int hashCode() {...3 lines}

        @Override
        public boolean equals(Object object) {...11 lines}

        @Override
        public String toString() {
            return String.format("%4d%4d%5.1f", user, item, rating);
        }
    }
}
```

Listing 9.18: Program to generate random ratings

(The folded code is the same as in the `DataGenerator1` program in *Listing 9.2*.) It creates random ratings from the set *{1.0, 1.5, 2.0, 2.5, ..., 5.0}* that are normally distributed with mean *3.0* and standard deviation *1.0*.

The only modifications needed for the `Filter` program are changing `int` to `double` where necessary. The results from a sample run are shown in *Figure 9.10* and *Figure 9.11*:

```
1  5 users
2  12 items
3   0.0  2.5  0.0  2.5  0.0  3.5  3.0  0.0  0.0  2.5  0.0  2.0
4   3.0  0.0  2.0  0.0  4.5  2.5  3.0  0.0  0.0  2.5  0.0  2.0
5   1.5  0.0  4.0  5.0  4.0  0.0  2.5  3.5  0.0  0.0  2.5  3.5
6   4.0  3.5  3.0  1.5  0.0  0.0  0.0  0.0  0.0  2.5  2.5  2.0
7   0.0  2.0  0.0  1.0  0.0  4.5  2.5  0.0  0.0  0.0  3.0  3.0
```

Figure 9.10: Utility file from DataGenerator3

```
1   12 items
2   1.00  0.57  0.85  0.44  0.62  0.23  0.44  0.29  0.00  0.77  0.57  0.64
3   0.57  1.00  0.41  0.48  0.00  0.60  0.48  0.00  0.00  0.73  0.67  0.66
4   0.85  0.41  1.00  0.77  0.77  0.15  0.54  0.74  0.00  0.54  0.70  0.77
5   0.44  0.48  0.77  1.00  0.57  0.36  0.69  0.85  0.00  0.39  0.71  0.84
6   0.62  0.00  0.77  0.57  1.00  0.30  0.71  0.66  0.00  0.43  0.36  0.66
7   0.23  0.60  0.15  0.36  0.30  1.00  0.85  0.00  0.00  0.56  0.47  0.71
8   0.44  0.48  0.54  0.69  0.71  0.85  1.00  0.45  0.00  0.63  0.54  0.89
9   0.29  0.00  0.74  0.85  0.66  0.00  0.45  1.00  0.00  0.00  0.54  0.61
10  0.00  0.00  0.00  0.00  0.00  0.00  0.00  0.00  0.00  0.00  0.00  0.00
11  0.77  0.73  0.54  0.39  0.43  0.56  0.63  0.00  0.00  1.00  0.31  0.60
12  0.57  0.67  0.70  0.71  0.36  0.47  0.54  0.54  0.00  0.31  1.00  0.85
13  0.64  0.66  0.77  0.84  0.66  0.71  0.89  0.61  0.00  0.60  0.85  1.00
```

Figure 9.11: Similarity file from DataGenerator3

Item #9 has not been bought by anyone — all five of its entries in the utility matrix are *0.0*. Consequently, in the similarity matrix, every entry in *row #9* or *column #9* is *0.0* — no item can be deemed to be at all similar to it.

Recommender Systems

To modify our `Recommender2` program to accommodate the decimal ratings, we must not only change `int` to `double` where necessary, but we must also adjust the `popularity()` method in the nested `Item` class. In `Recommender2`, we measured the popularity of the items simply by the number of users who had bought them. That worked because all ratings were the same value: one. But now, with different valued ratings, we should compute the average of the ratings for each item to determine its popularity.

The `getInput()` method for `Recommender3` now requires three (interactive) inputs: the current `user`, the `item` purchased by that user, and the `rating` that user gives to that item, as shown at lines 70-85 in *Listing 9.19*:

```java
public class Recommender3 {
    private static int m;               // number of users
    private static int n;               // number of items
    private static double[][] u;        // utility matrix
    private static double[][] s;        // similarity matrix
    private static int user;            // the current user
    private static Item itemBought;     // the current item bought by user

    public static void main(String[] args) {...7 lines }

    public static void readFiles() {...10 lines }

    public static void readUtilMatrix(File f) throws FileNotFoundException {...13 lines }

    public static void readSimilMatrix(File f) throws FileNotFoundException {...13 lines }

    public static void getInput() {
        Scanner input = new Scanner(System.in);
        System.out.print("Enter user number: ");
        user = input.nextInt();
        System.out.print("Enter item number: ");
        int item = input.nextInt();
        if (u[user][item] > 0) {
            System.out.printf("User %d already has item %d.%n", user, item);
            System.exit(0);
        }
        System.out.print("Enter rating (1-5): ");
        double rating = input.nextDouble();
        System.out.printf("User %d rated item %d at %4.2f.%n", user, item, rating);
        u[user][item] = rating;
        itemBought = new Item(item);
    }
```

Listing 9.19: The getInput() method for the Recommender3 program

```
121        static class Item {
122            int index;
123
124            public Item(int index) {...3 lines }
127
128            public double popularity() {
129                double sum = 0.0;
130                int count = 0;
131                for (int i = 1; i <= m; i++) {
132                    double value = u[i][this.index];
133                    if (value > 0) {
134                        sum += value;
135                        ++count;
136                    }
137                }
138                return (count > 0 ? sum/count : 0.0);
139            }
140
141            public double similarity(Item item) {...3 lines }
144
145            public class PopularityComparator implements Comparator<Item> {...8 lines }
153
154            public class SimilarityComparator implements Comparator<Item> {...8 lines }
162
163            @Override
             public String toString() {...3 lines }
167        }
```

Listing 9.20: Item class for the Recommender3 program

To see why we should use the average instead of the sum of the ratings to measure an item's preferability, suppose the first two columns of the utility matrix are (5.0, 4.0, 0.0, 0.0, 0.0) and (2.0, 2.0, 2.0, 2.0 1.5). The sums for the two items are 9.0 and 9.5; if we were to use that measure, then item two would be deemed preferable to item one. But the averages for the two items are 4.5 and 1.9, clearly indicating that item one is preferable. The question is, what should popular mean? This is really a software design question. Our answer is to use averages.

Figure 9.12 shows a sample run of the `Recommender3` program. *User #1* buys *item #1*, rating it 2.5. The set of other items not yet bought by that user is {3, 5, 8, 9, 11}. After ordering that set according to similarity with *item #1*, it is set1 = (3, 5, 11, 8, 9), with corresponding similarities 0.85, 0.62, 0.57, 0.29, and 0.0 (see *Figure 9.11*). The first four (n/3) of those are (3, 5, 11, 8). After reordering that set according to popularity, it becomes set2 = (5, 8, 3, 11), with corresponding popularities 4.25, 3.50, 3.00, and 2.67 (see *Figure 9.10*). The first three (n/4) of those are (5, 8, 3).

```
Output - RecommenderSystems (run)
run:
Enter user number: 1
Enter item number: 1
Enter rating (1-5): 2.5
User 1 rated item 1 at 2.50.
We also recommend these 3 items:  5  8  3
```

Figure 9.12: Run of Recommender3 Program

Large sparse matrices

In commercial implementations of these recommender systems, the utility and similarity matrices would be far too large to be stored as internal arrays. Amazon, for example, has millions of items for sale and hundreds of millions of customers. With m = 100,000,000 and n = 1,000,000, the utility matrix would have $m \, n$ = 100,000,000,000,000 slots and the similarity matrix would have n_2 = 1,000,000,000,000 slots. Moreover, if the average customer buys 100 items, then only $100n$ = 100,000,000 of the entries of the utility matrix would be non-zero—that's only 0.0001 percent of the entries, making it a very sparse matrix.

A **sparse matrix** is a matrix in which nearly all the entries are zero. Even if possible, it is very inefficient to store such a matrix as a two-dimensional array. In practice, other data structures are used.

There are several data structures that are good candidates for storing sparse matrices. A **map** is a data structure that implements a mathematical function $y = f(x)$. With a function, we think of the independent variable x as the input and the dependent variable y as the output. In the context of a map, the input variable x is called the key, and the output variable y is called the value.

With a mathematical function $y=f(x)$, the variables can be integers, real (decimal) numbers, vectors, or even more general mathematical objects. A one-dimensional array is a map where the key x ranges over the set of integers {0, 1, 2, . . . , n-1} (in Java). We write a[i] instead of a(i), but it's the same thing. Similarly, a two-dimensional array is a map where the key x ranges over a set of pairs of integers {(0,0), (0,1), (0,2),..., (1,0), (1,1), (1,2), . . . , (m,0), (m,1), (m,2), . . . , (m-1, n-1)}, and we write a[i][j] instead of $a((i, j))$.

Java implements the map data structure with the interface java.util.Map<K,V> and a large number of implementing classes (Java 8 provides 19 classes that implement the Map interface). The type parameters K and V stand for key and value. Probably the most commonly used implementations are the HashMap and TreeMap classes. The HashMap class is Java's standard implementation of the **hash table** data structure. The TreeMap class implements the **red-black tree** data structure, which maintains the order of the elements as determined by the ordering defined on the V class.

The SparseMatrix class in *Listing 9.21* illustrates an elementary implementation for sparse matrices. Its backing store is a Map<Key,Double> object, where Key is defined as an inner class. The advantage of using this SparseMatrix class instead of a two-dimensional array for the utility matrix is that it stores only the data from the Purchases.dat file – each element represents an actual (user, item) purchase.

Recall that in Java, an inner class is a non-static nested class. Our definition of the nested Key class here cannot be static because it accesses the non-static field n in its hashCode() method.

```java
SparseMatrix.java
11  public class SparseMatrix {
12      private final int m, n;   // dimensions of matrix
13      private final Map<Key,Double> map;
14
15      public SparseMatrix(int m, int n) {...5 lines}
20
21      public void put(int i, int j, double x) {
22          map.put(new Key(i,j), x);
23      }
24
25      public double get(int i, int j) {
26          return map.get(new Key(i,j));
27      }
28
29      public class Key implements Comparable {
30          int i, j;
31
32          public Key(int i, int j) {...4 lines}
36
37          @Override
38          public int hashCode() {
39              return i*n + j;
40          }
41
42          @Override
43          public boolean equals(Object object) {
44              if (object == null) {
45                  return false;
46              } else if (object == this) {
47                  return true;
48              } else if (!(object instanceof Key)) {
49                  return false;
50              }
51              Key that = (Key)object;
52              return that.i == this.i && that.j == this.j;
53          }
54
55          @Override
56          public String toString() {
57              return String.format("(%d,%d)", i, j);
58          }
59
60          @Override
61          public int compareTo(Object object) {
62              if (object == null) {
63                  return -1;
64              } else if (object == this) {
65                  return 0;
66              } else if (!(object instanceof Key)) {
67                  return -1;
68              }
69              Key that = (Key)object;
70              return this.hashCode() - that.hashCode();
71          }
72      }
73  }
```

Listing 9.22: A SparseMatrix Class

To use this, we make these changes to the code:

- Replace `double[] []` with `SparseMatrix` in declarations of the utility matrix u
- Replace the initialization `new double[m+1][n+1]` with `new SparseMatrix(m,n)`
- Replace references `u[i][j]` with `u.get(i,j)`
- Replace assignments `u[i][j] = x` with `u.put(i,j,x)`

This implementation sacrifices efficiency for simplicity. It is inefficient because it will generate a large number of objects: every entry is an object containing two other objects, and every call to `put()` and to `get()` generates another object.

A more efficient implementation would be to use a two-dimensional array of `int` for the keys and a one-dimensional array of `double` for the values, like this:

- `int[] [] key;`
- `double[] u;`

The value `key[i][j]` would be the index into the `u[]` array. In other words, u_{ij} would be stored as `u[key[i][i]]`. The `key[]` array would be maintained in **lexicographic order**.

Note that this class would not be a good implementation for the similarity matrix, because it is not sparse. But it is symmetric. Consequently, we can halve the work of the `computeSimilarityMatrix()` method with the improvements shown in Listing 9.22:

```java
    public static double[][] computeSimilarityMatrix(SparseMatrix u) {
        double[][] s = new double[n+1][n+1];
        for (int j = 1; j <= n; j++) {
            for (int k = 1; k < j; k++) {
                s[j][k] = s[k][j] = cosine(u, j, k);
            }
        }
        for (int j = 1; j <= n; j++) {
            s[j][j] = 1.0;
        }
        return s;
    }
```

Listing 9.22: A more efficient method for the similarity matrix computation

The double assignment at line 72 eliminates re-computing the duplicate `cosine(u, k, j)` value. And the loop at lines 75-77 assigns the correct value for all diagonal elements without using the `cosine()` method.

Using random access files

Another alternative for storing the similarity matrix (and the utility matrix) is to use a `RandomAccessFile` object. This is illustrated in *Listing 9.23*:

```java
public class RandomAccessFileTester {
    private static final int W = Double.BYTES;  // 8

    public static void main(String[] args) {
        String filespec = "data/Similarity4.dat";
        try {
            RandomAccessFile inout = new RandomAccessFile(filespec, "rw");
            for (int i = 0; i < 100; i++) {
                inout.writeDouble(Math.sqrt(i));
            }
            System.out.printf("Current file length is %d.%n", inout.length());

            for (int i = 4; i < 10; i++) {
                inout.seek(i*W);
                double x = inout.readDouble();
                System.out.printf("The square root of %1d is %.8f.%n", i, x);
            }
            System.out.println();

            inout.seek(7*W);
            inout.writeDouble(9.99999);

            for (int i = 4; i < 10; i++) {
                inout.seek(i*W);
                double x = inout.readDouble();
                System.out.printf("The square root of %1d is %.8f.%n", i, x);
            }
            inout.close();
        } catch (IOException e) {
            System.err.println(e);
        }
    }
}
```

```
run:
Current file length is 800.
The square root of 4 is 2.00000000.
The square root of 5 is 2.23606798.
The square root of 6 is 2.44948974.
The square root of 7 is 2.64575131.
The square root of 8 is 2.82842712.
The square root of 9 is 3.00000000.

The square root of 4 is 2.00000000.
The square root of 5 is 2.23606798.
The square root of 6 is 2.44948974.
The square root of 7 is 9.99999000.
The square root of 8 is 2.82842712.
The square root of 9 is 3.00000000.
BUILD SUCCESSFUL (total time: 0 seconds)
```

Listing 9.23: Using a random access file

This little test program creates a random access file named `inout` at line 17. The constant `W`, defined at line 12, is the number of bytes (8) that Java uses to store a `double` value. We need that to locate our data in the file. The second argument to the constructor, the string `rw`, means that we will be both reading from and writing to the file. The loop at lines 18-20 writes 100 square roots into the file. The output from line 21 confirms that the file contains 800 bytes.

The loop at lines 23-27 uses direct access (random access) into the file, just like accessing a 100-element array. Each access requires two steps: seek the location to be read, and then read it. The `seek()` method sets the file's read-write pointer to the point in the file where access is to begin. Its argument is the byte address of that starting point; it is called the **offset** – the distance from the beginning of the file.

The loop at lines 23-27 iterates six times, for i = 4 through 9. On iteration i, it reads eight bits (one byte), from bit $8i$ through bit $8i + 7$. For example, when $i = 4$, it reads bits 32-39; that chunk of bits has offset 32. The byte that is read each time is stored in x, as a `double` value. The print statement at line 26 prints that number.

The statements at lines 30-31 illustrate how to change a value stored in the file. The `seek()` method moves the read-write pointer to byte number 56 (that's 7*8); then the `writeDouble()` method overwrites the next eight bytes with the code that represents the number 9.99999. The loop at lines 33-37 shows that the number 2.64575131 has been replaced by that number.

That code is logically equivalent to this:

```
Double a = new double[100];
for (int i = 4; i < 10; i++) {
    a[i] = Math.sqrt(i);
}
a[7] = 9.99999;
```

In this respect, a random access file is equivalent to an array. But of course, a file can be much larger than an array.

The program in *Listing 9.23* illustrates how to store and access the equivalent of a one-dimensional array in a random access file. But the similarity array is two-dimensional. One might think that would make it a lot more complicated. But in fact, a simple calculation is all we need to flatten two dimensions into one. Actually, we already have done that in our `SparseMatrix` implementation. At line 39 in *Listing 9.21*, we used the expression `i*n + j` for the `hashCode()` in the inner `Key` class. So, for example, the hash code for the `key (2,7)` would be 2*10 + 7 = 27, assuming n = 10. This lexicographic ordering linearizes the two-dimensional array.

A random access file is not a text file.

It cannot be read in an ordinary text editor. But you can read and print chunks of it, using the `read` methods that are provided by the `RandomAccessFile` class. There are 17 of them, including `readDouble()` and `readLine()`.

The code in *Listing 9-24* shows how we can compute and store the similarity matrix s in a random access file:

```
68      public static void computeSimilarityMatrix(SparseMatrix u,
69              RandomAccessFile s) throws IOException {
70          for (int j = 1; j <= n; j++) {
71              for (int k = 1; k < j; k++) {
72                  double x = cosine(u, j, k);
73                  s.seek((j*n + k - n - 1)*W);
74                  s.writeDouble(x);
75                  s.seek((j + k*n - n - 1)*W);
76                  s.writeDouble(x);
77              }
78          }
79          for (int j = 1; j <= n; j++) {
80              s.seek((j - 1)*(n + 1)*W);
81              s.writeDouble(1.0);
82          }
83      }
```

Listing 9.24: Storing similarities in a random access file

The arguments to the `seek()` method at lines 73, 75, and 80 are offsets. The formulas are a little tricky (and not essential to the general understanding here) because the similarity matrix uses one-based indexing. They are presented here mainly to demonstrate how a sparse similarity matrix could be stored in a random access file.

The Netflix prize

In 2006, Netflix announced that it would award a $1,000,000 prize to the best recommender algorithm submitted that could outperform their own algorithm. Two years later, the prize was awarded to a team called BellKor for their Pragmatic Chaos system. Netflix never used the prize-winning Pragmatic Chaos system, explaining that a production version would be too expensive to implement. That prizewinner turned out to be a mix of over 100 different methods. Meanwhile, some of the top competitors went on to extend and market their own recommender systems. Some of the resulting algorithms have been patented.

The competition was open to anyone who registered. Data for testing proposed algorithms was provided by Netflix. The main dataset was a list of 100,480,507 triples: a user ID number, a movie ID number, and a rating number from 1 to 5. The data included over 480,000 customer IDs and over 17,000 movie IDs. That's a very large utility matrix, which is also very sparse: about 99 percent empty.

Netflix's own recommender system generates about 30 billion predictions per day. The prize's requirement was that the winning system had to beat that system by at least 10%. Over 50,000 teams participated.

You can still download the data from the `http://academictorrents.com/` website. It is over 2 GB.

Summary

In this chapter, we have described the general strategy of recommender systems and implemented in Java an early version developed at Amazon. We first explored the notion of similarity measures, including cosine similarity. We saw how user ratings are used in recommender systems. We looked at the general idea of sparse matrices, which is the likely mathematical structure for a utility matrix, and then saw how they could be implemented using a random access file. Finally, we reviewed the Netflix prize competition, which raised the level of interest in recommender systems among data scientists.

10
NoSQL Databases

In *Chapter 5, Relational Databases*, we reviewed relational databases and the SQL query language used for processing them. Remember that data in a relational database is stored in tables, which are structured datasets.

In many modern software environments, data is too fluid, dynamic, and large to benefit from the constraints of a tightly designed relational database. In such cases, a non-relational database is preferred. Since the SQL query language does not apply to such storage arrangements, these non-relational databases are called **NoSQL databases**. They are particularly popular in successful web-based companies, such as Facebook, Amazon, and Google. To understand how data is managed and analyzed in these environments, we need to understand how the NoSQL databases themselves work.

The Map data structure

The underlying data structure for a NoSQL database is really the same as that for a relational database: the Map data structure (also called a **dictionary** or **associative array**), implemented in Java by the `java.util.Map<K,V>` interface and the 19 classes that implement it. Among them is the `HashMap<K,V>` class, which implements the classic hash table data structure. The type parameters K and V stand for key and value.

As described in *Chapter 9, Recommender Systems* the essential feature of the map data structure is its functional key-value mechanism. Like a mathematical function $y=f(x)$, the key-value mechanism is an input-output process. In the mathematical context, x is the input and y is the output. In the data structure context, the key is the input and the value is the output.

NoSQL Databases

Note that with a mathematical function, each *x*-value corresponds to only one *y*-value; you cannot have *f(7)* = 12 and *f(7)* = 16. In other words, the input values must be unique. In this way, each *x* value can be thought of as a (unique) identifier for the *y*-value, much like student IDs assigned to students at a university.

A good example of this mechanism is the index of a book. The keys are the key words, listed alphabetically in the index and the value for each key is the list of page numbers where that key word can be found. Note that each word appears only once in an index.

In a relational database table, one or more attributes (columns) are designated to be the key for the table. In the simplest case, one attribute would be the key. In that case:

- All entries in the key column must be unique (no duplicates)
- The vector of all the other entries in a row constitute the value for that row's key
- The database system provides an efficient indexing mechanism for rapidly searching for keys

Table 10-1 reproduces *Table 5-1* from *Chapter 5, Relational Databases*.

ID	Last Name	First Name	Date of Birth	Job Title	Email
49103	Adams	Jane	1975-09-02	CEO	jadams@xyz.com
15584	Baker	John	1991-03-17	Data Analyst	jbaker@xyz.com
34953	Cohen	Adam	1978-11-24	HR Director	acohen@xyz.com
23098	Davis	Rose	1983-05-12	IT Manager	rdavis@xyz.com
83822	Evans	Sara	1992-10-10	Data Analyst	sevans@xyz.com

Table 10-1. A table in a relational database

Here, the key attribute is *ID*. For example, the *ID* number 23098 identifies the employee *Rose Davis*. The value for that key is the vector of the other five fields in that row; that is, the vector (`"Davis"`, `"Rose"`, `"1983-05-12"`, `IT Manager`, `rdavis@xyz.com`). With the number 23098 as input, that vector would be the output.

The little program in *Listing 10-1* shows how we could implement this table in Java. The `HashMap` object `map` is declared at line 13 to have type `Map<Integer,Employee>`, and with an initial capacity of 100. The nested `Employee` class is defined at lines 20-31: it's instances are simple vectors of five strings, corresponding to the five other columns in *Table 10-1*. The code at lines 15-17 insert the record for `Rose Davis` into the `map`. The key is the number 23098, and the value is the `rose` object that is instantiated at line 15. Note that the `put()` method takes two arguments: the key and the value.

Most things we buy these days have product ID numbers. For example, books are identified by two different **International Standard Book Numbers, ISBN-10** and **ISBN-13** (see *Figure 10-1*), and individual cars are identified by their 17-character **vehicle identification number (VIN)**.

```
Example1.java
11  public class Example1 {
12      public static void main(String[] args) {
13          Map<Integer,Employee> map = new HashMap(100);
14
15          Employee rose = new Employee("Davis", "Rose", "1983-05-12",
16                  "IT Manager", "rdavis@xyz.com");
17          map.put(23098, rose);
18      }
19
20      static class Employee {
21          String lastName, firstName, dob, title, email;
22
23          public Employee(String lastName, String firstName, String dob,
24                  String title, String email) {
25              this.lastName = lastName;
26              this.firstName = firstName;
27              this.dob = dob;
28              this.title = title;
29              this.email = email;
30          }
31      }
32  }
```

Listing 10-1. Map implementation of the data structure in Table 10-1

Figure 10-1. Buying books by their ISBN identifiers

NoSQL Databases

Brick-and-mortar stores like Walmart use 12-digit **Universal Product Codes** (**UPCs**), read by bar-codes, to identify all their products and Amazon has its own **Amazon Standard Identification Number** (**AISM**).

Even things we don't buy have keys (unique identifiers). The URL for a webpage is its key.

IDs are keys, each matched with a value that could be a string, a vector, or an external file. The key-value paradigm is ubiquitous. So, it is natural that it should be the basis for databases, both SQL and NoSQL.

SQL versus NoSQL

Databases are generalized data structures. Both store data, either internally in memory or externally on disk or in the cloud. As data containers, they have a logical structure and a physical structure.

Consider the simplest of data structures: a one-dimensional array a[] of strings. The logical structure of this is shown in *Figure 10-2*.

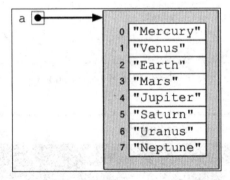

Figure 10-2. An array of strings

It is an object, referenced by the variable a. Inside that object is a sequence of numbered storage compartments, each capable of holding a string object.

However, the physical structure, hidden from the programmer, is a sequence of bytes in memory. Using two-byte Unicode characters, it will allocate 16 bytes for the encodings of the characters of the eight string, and it will also store information, such as the name of the array (a), the datatype of the elements being stored (String), and the hexadecimal starting location of the sequence of 16 bytes, elsewhere.

The same dichotomy holds for database structures, except that the actual storage is on disk (or the cloud) and its complexity is a magnitude greater. Fortunately, the software designers and engineers can imagine the logical structure most of the time.

As discussed in *Chapter 5, Relational Databases*, the logical structure of a relational database is a collection of tables and associated links. These are maintained by the database system, which is controlled by programs mostly written in the SQL query language.

A NoSQL database does not use tables to store its data. Its logical structure can be imagined as a large collection of key-value maps, each stored as a separate document. As we have just seen, this is similar to relational tables with specified key attributes. But, without the use of SQL, the database is less lightly structured. The trade-off is that, for the kind of operations needed for web-based software, the system will be more flexible and efficient, especially with very large datasets.

Relational databases and the SQL language were developed in the 1970s. They became the standard database environment for the secure management of stable institutional data. However, with the advance of Web commerce, the demand for data management shifted: datasets became much larger and much more dynamic. NoSQL database systems responded to that shift in requirements.

There are several popular NoSQL database systems in use, including MongoDB, Cassandra, HBase, and Oracle NoSQL Database. We shall examine MongoDB in the following sections.

The Mongo database system

The Mongo database system, MongoDB, has been in development since 2007. It has become the most widely used NoSQL database system. The name is a substring of the word humongous, suggesting that it works well with very large datasets (it has also been suggested that the name is after a character, played by the great NFL defensive tackle, Alex Karras, in the Mel Brooks movie, *Blazing Saddles*). The system is described as a document-oriented database.

You can download MongoDB from `https://www.mongodb.com/download-center`.

See the *Appendix* for details on installing MongoDB.

NoSQL Databases

After installing MongoDB, start the database system with the `mongo` command, as shown in *Figure 10-3*. We are using the command line here. This is done using the Terminal app on a Mac, the Command Prompt on a PC, and a Shell window on a UNIX box.

```
~ $ mongod
2017-07-25T09:37:14.402-0400 I CONTROL  [initandlisten] MongoDB starting : pid=6481 port=27017 dbpath=/data/db 64-bit
2017-07-25T09:37:14.403-0400 I CONTROL  [initandlisten] db version v3.4.6
2017-07-25T09:37:14.403-0400 I CONTROL  [initandlisten] git version: c55eb86ef46ee7aede3b1e2a5d184a7df4bfb5b5
2017-07-25T09:37:14.403-0400 I CONTROL  [initandlisten] OpenSSL version: OpenSSL 0.9.8zh 14 Jan 2016
2017-07-25T09:37:14.403-0400 I CONTROL  [initandlisten] allocator: system
2017-07-25T09:37:14.403-0400 I CONTROL  [initandlisten] modules: none
2017-07-25T09:37:14.403-0400 I CONTROL  [initandlisten] build environment:
2017-07-25T09:37:14.403-0400 I CONTROL  [initandlisten]     distarch: x86_64
```

Figure 10-3. Starting the Mongo database system from the command line

The output continues for many more lines, and then it pauses.

Set that command window aside, open a new one, and then execute the `mongo` command, as shown in *Figure 10-4*:

```
~ $ mongo
MongoDB shell version v3.4.6
connecting to: mongodb://127.0.0.1:27017
MongoDB server version: 3.4.6
Server has startup warnings:
2017-07-25T09:37:18.211-0400 I CONTROL  [initandlisten]
2017-07-25T09:37:18.211-0400 I CONTROL  [initandlisten] ** WARNING:
2017-07-25T09:37:18.211-0400 I CONTROL  [initandlisten] **
2017-07-25T09:37:18.211-0400 I CONTROL  [initandlisten]
2017-07-25T09:37:18.211-0400 I CONTROL  [initandlisten]
2017-07-25T09:37:18.211-0400 I CONTROL  [initandlisten] ** WARNING:
>
```

Figure 10-4. Starting the MongoDB shell from the command line

The `mongo` command starts up MongoDB shell, allowing the execution of MongoDB commands. Notice that the command prompt changes to an angle bracket >.

Notice the two startup warnings. The first refers to access control and can be remedied by defining an administrative user with a password. For details, go to:

https://docs.mongodb.com/master/tutorial/enable-authentication/.

The second warning refers to your operating system's user limit on the number of files that can be open at a time, 256 files in this case. This page refers to that:

https://docs.mongodb.com/manual/reference/ulimit/.

```
> show dbs
admin   0.000GB
local   0.000GB
> use friends
switched to db friends
> tom = {fname:"Tom", lname:"Jones", dob:"1983-07-25", sex:"M"}
{ "fname" : "Tom", "lname" : "Jones", "dob" : "1983-07-25", "sex" : "M" }
> tom
{ "fname" : "Tom", "lname" : "Jones", "dob" : "1983-07-25", "sex" : "M" }
> db.friends.insert(tom);
WriteResult({ "nInserted" : 1 })
> ann = {fname:"Ann", lname:"Smith", dob:"1986-04-19", sex:"F"}
{ "fname" : "Ann", "lname" : "Smith", "dob" : "1986-04-19", "sex" : "F" }
> jim = {fname:"Jim", lname:"Chang", dob:"1986-04-19", sex:"M"}
{ "fname" : "Jim", "lname" : "Chang", "dob" : "1986-04-19", "sex" : "M" }
> db.friends.insert(ann);
WriteResult({ "nInserted" : 1 })
> db.friends.insert(jim);
WriteResult({ "nInserted" : 1 })
> show dbs
admin    0.000GB
friends  0.000GB
local    0.000GB
>
```

Figure 10-5. *Creating a database in MongoDB*

The commands executed in *Figure 10-5* create a MongoDB database with a collection of three documents. The name of the database is `friends`.

The `show dbs` command at first shows only two databases in the system: the default databases, `admin` and `local`. The `use friends` command creates the `friends` database. Next, we create three documents, named `tom`, `ann`, and `jim`, and add them to the database with the `db.friends.insert()` statement. Note that, like Java, MongoDB shell statements must be terminated with a semicolon. The last `show dbs` command confirms that the friends database has been created.

NoSQL Databases

The `find()` command is used in *Figure 10-6*.

```
> db.friends.find()
{ "_id" : ObjectId("597772eb142e364c7dff8681"), "fname" : "Tom", "lname" : "Jones",
"dob" : "1983-07-25", "sex" : "M" }
{ "_id" : ObjectId("59777306142e364c7dff8682"), "fname" : "Ann", "lname" : "Smith",
"dob" : "1986-04-19", "sex" : "F" }
{ "_id" : ObjectId("5977730e142e364c7dff8683"), "fname" : "Jim", "lname" : "Chang",
"dob" : "1986-04-19", "sex" : "M" }
> oid = ObjectId("597772eb142e364c7dff8681")
ObjectId("597772eb142e364c7dff8681")
> oid.getTimestamp()
ISODate("2017-07-25T16:33:47Z")
>
```

Figure 10-6. Using the MongoDB find() and getTimestamp() methods

It lists the collection in the `friends` database. Note that each of the three documents has been given an `ObjectID` for its `_id` field. This object contains a timestamp of the instant when the document was created and it identifies the machine and process that created it. As shown, the `getTimestamp()` command shows the timestamp stored in the referenced object.

NoSQL databases are structurally different from **relational databases** (**Rdbs**) in many fundamental ways, but logically, there is some correspondence with their data structures. This is summarized in *Table 10-2*.

Relational Database	MongoDB
database	database
table (relation)	collection
row (record, tuple)	document
entry (field, element)	key-value pair

Table 10-2. Rdb and MongoDB data structures

A relational database can be imagined as mainly a set of tables; similarly, a MongoDB database is a set of collections. An Rdb table is a set of rows, each adhering to the datatype schema defined for that table. Analogously, a collection is a set of documents, each document being stored as a binary JSON file (a BSON file; **JSON (JavaScript Object Notation**) files are discussed in *Chapter 2, Data Preprocessing*). Finally, an Rdb table row is a sequence of data entries, one for each column, while a MongoDB document is a set of key-value pairs.

This data model is known as a **document store**, and is used by some of the other leading NoSQL database systems, such as IBM Domino and Apache CouchDB. In contrast, Apache Cassandra and HBase use the **column store** data model, where a column is defined to be a key-value pair with a timestamp.

Notice how much freer the data design process is with a NoSQL database as opposed to an Rdb. The Rdb requires a rather tightly defined preliminary data architecture that specifies tables, schema, datatypes, keys, and foreign keys before we can even create the tables. The only preliminary data design needed with MongoDB, before we begin inserting data, is the definitions of the collections.

In an Rdb w, if z is a data value for column y in table x, then x can be referenced as $w.x.y.z$; that is, the names of the database, the table, and the column can serve as namespace identifiers. In contrast, the corresponding namespace identifier in a MongoDB would be $w.x.y.k.z$, where x is the name of the collection, y is the name of the document, and k is the key in the key-value pair for x.

We saw previously that the MongoDB shell will automatically create a collection with the same name as the database itself if we use the command `db.name.insert()`; so, we already have a collection named `friends`. The code in *Figure 10-7* illustrates how to create a collection explicitly, in this case one named `relatives`:

```
> db.createCollection("relatives")
{ "ok" : 1 }
> show collections
friends
relatives
> db.relatives.insert({'fname':'Jack','relation':'grandson'})
WriteResult({ "nInserted" : 1 })
> db.relatives.insert([
... {'fname':'Henry','relation':'grandson'},
... {'fname':'Sara','relation':'daughter'}
... ])
BulkWriteResult({
        "writeErrors" : [ ],
        "writeConcernErrors" : [ ],
        "nInserted" : 2,
        "nUpserted" : 0,
        "nMatched" : 0,
        "nModified" : 0,
        "nRemoved" : 0,
        "upserted" : [ ]
})
> db.relatives.find()
{ "_id" : ObjectId("597798d5142e364c7dff8687"), "fname" : "Jack", "relation" : "grandson" }
{ "_id" : ObjectId("59779902142e364c7dff8688"), "fname" : "Henry", "relation" : "grandson" }
{ "_id" : ObjectId("59779902142e364c7dff8689"), "fname" : "Sara", "relation" : "daughter" }
>
```

Figure 10-7. Creating a separate MongoDB collection

The `show collections` command shows that we have those two collections now.

Next, we insert three documents into our `relatives` collection. Notice the following features:

- We can use either the double quote character (")or the single quote character (') to delimit strings
- We can spread the command out over several lines, provided we get all the punctuation right, matching quotes, braces, parentheses, and brackets correctly
- The `insert` command can insert a list of documents, using the bracket characters (`[]`) to delimit the list. This is called a **bulk write**

> For longer commands, it helps to write them out first in a text editor and then copy it to the command line. If you do, be sure to use the right quote characters.

The next example, shown in *Figure 10-8*, illustrates a compound query:

```
> db.friends.find({$and:[{'sex':'M'},{'dob':{$gt:'1985-01-01'}}]}).pretty()
{
         "_id"   : ObjectId("5977730e142e364c7dff8683"),
         "fname" : "Jim",
         "lname" : "Chang",
         "dob"   : "1986-04-19",
         "sex"   : "M"
}
```

Figure 10-8. A MongoDB compound query

The argument to the `find()` command contains a conjunction of two conditions. The conjunction is the logical AND operator, written as `&&` in Java and as `$and` in MongoDB. In this example, the two conditions are that `'sex'` is `'M'` and `'dob'` > `'1985-01-01'`. In other words, "Find all male friends who were born after January 1, 1985".

The appended `pretty()` method simply tells the shell to use multi-line formatting for the results.

In MongoDB, the two logical operators AND and OR are written as `$and:` and `$or:`. The six arithmetic operators, <, ≤, >, ≥, ≠, and =, are written as `$lt:`, `$lte:`, `$gt:`, `$gte:`, `$ne:`, and :.

After getting used to the syntax, you could almost guess the right form for the `update` operation. It is illustrated in *Figure 10-9*:

```
> db.friends.update({'fname':'Tom'},{$set:{'phone':'123-456-7890'}})
WriteResult({ "nMatched" : 1, "nUpserted" : 0, "nModified" : 1 })
> db.friends.find().pretty()
{
        "_id" : ObjectId("597772eb142e364c7dff8681"),
        "fname" : "Tom",
        "lname" : "Jones",
        "dob" : "1983-07-25",
        "sex" : "M",
        "phone" : "123-456-7890"
}
{
        "_id" : ObjectId("59777306142e364c7dff8682"),
        "fname" : "Ann",
        "lname" : "Smith",
        "dob" : "1986-04-19",
        "sex" : "F"
}
{
        "_id" : ObjectId("5977730e142e364c7dff8683"),
        "fname" : "Jim",
        "lname" : "Chang",
        "dob" : "1986-04-19",
        "sex" : "M"
}
```

Figure 10-9. Using the MongoDB update() method

Here we have added a phone number to our friend Tom's document.

You can use the `update()` method to change existing fields or to add a new one.

Notice that this method will change all the documents whose existing data satisfy the first argument. If we had used `{'sex'='M'}` instead of `{'fname'='Tom'}`, that phone number would have been added to both documents whose `sex` field is M.

When you have finished using `mongo`, execute the `quit()` method to terminate the session and return to the OS command line. This is illustrated in *Figure 10-10*:

```
Last login: Wed Jul 26 07:48:39 on ttys003
[~ $ mongo
MongoDB shell version v3.4.6
connecting to: mongodb://127.0.0.1:27017
MongoDB server version: 3.4.6
Server has startup warnings:
2017-07-25T09:37:18.211-0400 I CONTROL  [initandlisten]
2017-07-25T09:37:18.211-0400 I CONTROL  [initandlisten] ** WARNING:
ol is not enabled for the database.
2017-07-25T09:37:18.211-0400 I CONTROL  [initandlisten] **
te access to data and configuration is unrestricted.
2017-07-25T09:37:18.211-0400 I CONTROL  [initandlisten]
2017-07-25T09:37:18.211-0400 I CONTROL  [initandlisten]
2017-07-25T09:37:18.211-0400 I CONTROL  [initandlisten] ** WARNING:
  too low. Number of files is 256, should be at least 1000
[> show dbs
admin         0.000GB
friends       0.000GB
local         0.000GB
peoplepedia   0.000GB
[> use friends
switched to db friends
[> show collections
friends
relatives
[> quit()
~ $
```

Figure 10-10. A complete MongoDB shell session

The online MongoDB Manual at `https://docs.mongodb.com/manual/introduction/` is an excellent source of examples and further information.

The Library database

In *Chapter 5, Relational Databases*, we created a `Library` database as an Rdb, using NetBeans Java DB relational database system. The design for that database is shown in *Figure 5-2*. (The same database could have been built using MySQL or any other Rdb.) Here, we will build a MongoDB database for the same data.

As mentioned previously, the only preliminary design decisions that we have to make are the names of the database itself and its collections. We'll name the database `library`, and its three collections `authors`, `publishers`, and `books`. These are created in *Figure 10-11*:

johnrhubbard — mongo — 72×11

```
> use library
switched to db library
>
> db.createCollection('authors')
{ "ok" : 1 }
> db.createCollection('publishers')
{ "ok" : 1 }
> db.createCollection('books')
{ "ok" : 1 }
>
```

Figure 10-11. Creating a library database

Then, we can insert some date, as shown in *Figure 10-12*:

johnrhubbard — mongo — 102×45

```
> db.authors.insert({'_id':'AhoAV','lname':'Aho','fname':'Alfred V.','yob':1941})
WriteResult({ "nInserted" : 1 })
> db.authors.insert({'_id':'HopcrofttJE','lname':'Hopcroft','fname':'John E.','yob':1939})
WriteResult({ "nInserted" : 1 })
> db.authors.insert({'_id':'WirthN','lname':'Wirth','fname':'Niklaus','yob':1934})
WriteResult({ "nInserted" : 1 })
> db.authors.insert({'_id':'LeisersonCE','lname':'Leiserson','fname':'Charles E.','yob':1953})
WriteResult({ "nInserted" : 1 })
> db.authors.insert({'_id':'RivestRL','lname':'Rivest','fname':'Ronald L.','yob':1947})
WriteResult({ "nInserted" : 1 })
> db.authors.insert({'_id':'SteinCL','lname':'Stein','fname':'Clifford S.','yob':1965})
WriteResult({ "nInserted" : 1 })
```

Figure 10-12. Inserting documents into the authors collection

Here, we have inserted six documents in the `authors` collection, representing six authors.

Notice that we have given each document four fields: `_id`, `lname`, `fname`, and `yob`. The first three are strings and the last is an integer. The `_id` field values combine the author's last name with first and middle initials.

Next, check the results:

```
> db.authors.find().sort({'_id':1})
{ "_id" : "AhoAV", "lname" : "Aho", "fname" : "Alfred V.", "yob" : 1941 }
{ "_id" : "HopcrofttJE", "lname" : "Hopcroft", "fname" : "John E.", "yob" : 1939 }
{ "_id" : "LeisersonCE", "lname" : "Leiserson", "fname" : "Charles E.", "yob" : 1953 }
{ "_id" : "RivestRL", "lname" : "Rivest", "fname" : "Ronald L.", "yob" : 1947 }
{ "_id" : "SteinCL", "lname" : "Stein", "fname" : "Clifford S.", "yob" : 1965 }
{ "_id" : "WirthN", "lname" : "Wirth", "fname" : "Niklaus", "yob" : 1934 }
```

Figure 10-13. *Examining the contents of the* authors *collection*

We used the `sort()` method here to have the documents output alphabetically by their `_id` values. The `1` is an argument to the `sort()` method asking for ascending order; `-1` would be in descending order.

Next, we insert four documents into the `publishers` collection and check the results.

Note that this screen capture is incomplete—it had to be truncated at the right margin:

```
> db.publishers.insert({'_id':'PACKT','name':'Packt Publishers Limited','city':'Birmingham','country':'UK',
WriteResult({ "nInserted" : 1 })
> db.publishers.insert({'_id':'MIT','name':'The MIT Press','city':'Cambridge, MA','country':'US','url':'mit
WriteResult({ "nInserted" : 1 })
> db.publishers.insert({'_id':'A-W','name':'Addison-Wesley Longman, Inc.','city':'Reading, MA','country':'U
WriteResult({ "nInserted" : 1 })
> db.publishers.insert({'_id':'PH','name':'Prentice Hall, Inc.','city':'Upper Saddle River, NJ','country':'
WriteResult({ "nInserted" : 1 })
>
> db.publishers.find().sort({'_id':1})
{ "_id" : "A-W", "name" : "Addison-Wesley Longman, Inc.", "city" : "Reading, MA", "country" : "US", "url" :
{ "_id" : "MIT", "name" : "The MIT Press", "city" : "Cambridge, MA", "country" : "US", "url" : "mitpress.mi
{ "_id" : "PACKT", "name" : "Packt Publishers Limited", "city" : "Birmingham", "country" : "UK", "url" : "p
{ "_id" : "PH", "name" : "Prentice Hall, Inc.", "city" : "Upper Saddle River, NJ", "country" : "US", "url"
>
```

Figure 10-14. *Inserting documents into the* publishers *collection*

We also insert documents into the `books` collection:

```
> db.books.insert({'_id':'9781491901632','title':'Hadoop: The Definitive Guide','author':'WhiteT','pub
WriteResult({ "nInserted" : 1 })
> db.books.insert({'_id':'9781449344689','title':'MongoDB: The Definitive Guide','author':'ChodorowK',
WriteResult({ "nInserted" : 1 })
> db.books.insert({'_id':'0201000237','title':'Algorithms and Data Structures','author':['AhoAV','Hopc
WriteResult({ "nInserted" : 1 })
> db.books.find().pretty()
{
        "_id" : "9781491901632",
        "title" : "Hadoop: The Definitive Guide",
        "author" : "WhiteT",
        "publisher" : "OREILLY",
        "year" : 2015
}
{
        "_id" : "9781449344689",
        "title" : "MongoDB: The Definitive Guide",
        "author" : "ChodorowK",
        "publisher" : "OREILLY",
        "year" : 2013
}
{
        "_id" : "0201000237",
        "title" : "Algorithms and Data Structures",
        "author" : [
                "AhoAV",
                "HopcroftJE",
                "UllmanJD"
        ],
        "publisher" : "A-W",
        "year" : 1982
}
```

Figure 10-15. *Inserting documents into the books collection*

Notice that for the third book, we have used an array object for the value of the `author` key:

```
"author " : ["AhoAV ", "HopcroftJE ", "UllmanJD " ]
```

This is like the Java syntax:

```
String[] authors = {"AhoAV ", "HopcroftJE ", "UllmanJD "}
```

Also note that, unlike relational databases with foreign keys, in MongoDB the referent need not have been inserted prior to its reference. The author key `"WhiteT"` is referenced in the first `insert` statement in *Figure 10-15*, even though we have not yet inserted a document with that referent in the `authors` collection.

It is apparent that loading NoSQL database collections this way, using separate `insert()` calls on the command line, is time consuming and error prone. A better approach is to do bulk inserts within a program, as explained in the next section.

Java development with MongoDB

To access a MongoDB database from a Java program, you must first download the mongo-java-driver JAR files. You can get them from here:

http://central.maven.org/maven2/org/mongodb/mongo-java-driver/

Choose a recent, stable version; like, 3.4.2. Download the two JAR files: `mongo-java-driver-3.4.2.jar` and `mongo-java-driver-3.4.2-javadoc.jar`.

The program in *Listing 10-2* shows how to use Java to access a Mongo database, in this case our `friends` database:

```java
public class PrintMongoDB {
    public static void main(String[] args) {
        MongoClient client = new MongoClient("localhost", 27017);
        MongoDatabase friends = client.getDatabase("friends");
        MongoCollection relatives = friends.getCollection("relatives");

        Bson bson = Sorts.ascending("fname");
        FindIterable<Document> docs = relatives.find().sort(bson);
        int num = 0;
        for (Document doc : docs) {
            String name = doc.getString("fname");
            String relation = doc.getString("relation");
            System.out.printf("%4d. %s, %s%n", ++num, name, relation);
        }
    }
}
```

```
1. Henry, grandson
2. Jack, grandson
3. Sara, daughter
```

Listing 10-2. Java program to print collection documents

At lines 18-20, we instantiate the three objects necessary to access our `relatives` collection: a `MongoClient` object, a `MongoDatabase` object, and a `MongoCollection` object. The rest of the program iterated through all the documents in that collection, numbering and printing them in alphabetical order.

Line 23 instantiates a specialized `Iterable` object that uses the `find()` method on the `relatives` collection, accessing its documents according to the ordering specified by the `bson` object that is defined at line 22. The loop at lines 25-29 then prints the two fields of each document.

This is the data that we inserted in *Figure 10-7*.

> A BSON object is a JSON object (see *Chapter 2, Data Preprocessing,*) in a binary-encoded format, which is likely to have faster access.

The program in *Listing 10-3* inserts three new documents into our `relatives` collection and then prints the entire collection:

```java
public class InsertMongoDB {
    public static void main(String[] args) {
        MongoClient client = new MongoClient("localhost", 27017);
        MongoDatabase friends = client.getDatabase("friends");
        MongoCollection relatives = friends.getCollection("relatives");

        addDoc("John", "son", relatives);
        addDoc("Bill", "brother", relatives);
        addDoc("Helen", "grandmother", relatives);

        printCollection(relatives);
    }

    public static void addDoc(String fname, String relation,
            MongoCollection collection) {
        Document doc = new Document();
        doc.put("fname", fname);
        doc.put("relation", relation);
        collection.insertOne(doc);
    }

    public static void printCollection(MongoCollection collection) {
        Bson bson = Sorts.ascending("fname");
        FindIterable<Document> docs = collection.find().sort(bson);
        int num = 0;
        for (Document doc : docs) {
            String name = doc.getString("fname");
            String relation = doc.getString("relation");
            System.out.printf("%4d. %s, %s%n", ++num, name, relation);
        }
    }
}
```

```
Output - NoSQLDatabases (run)
   1. Bill, brother
   2. Helen, grandmother
   3. Henry, grandson
   4. Jack, grandson
   5. John, son
   6. Sara, daughter
```

Listing 10-3. Java program to insert documents into a collection

Each insertion is managed by the method defined at lines 29-35. This method adds a new document, representing another relative, to the collection.

The printing is done by the method at lines 37-46. That is the same code as at lines 22-30 in *Listing 10-2*.

Notice that the three new documents are included in the output, which is sorted by the `fname` field.

Keep in mind that a `MongoDatabase` is a set of `MongoCollection` objects, a `MongoCollection` is a set of `Document` objects, and a `Document` is a Map; that is, set of key-value pairs. For example, the first document accessed at line 24 is the key-value pair `{"fname":"Jack"}`.

A NoSQL database is logically similar to a **relational database (Rdb)**, in that a collection of key-value pairs is similar to an Rdb table. We could store the collection of the six key-value pairs from the program in *Listing 10-2* as an Rdb table, like the one shown in *Table 10-3*. Each document corresponds to one row in the table.

fname	relation
Bill	brother
Helen	grandmother
Henry	grandson
Jack	grandson
John	son
Sara	daughter

Table 10-3. Key-value pairs

However, a NoSQL collection is more general than the corresponding Rdb table. The documents in a collection are independent, with different numbers of elements and completely different fields and datatypes. However, in a table, every row has the same structure: the same number of fields (columns) and the same sequence of datatypes (schema).

Remember from *Chapter 5, Relational Databases*, that an Rdb table schema is a sequence of datatypes for the corresponding sequence of columns in the table. Each table has a unique schema to which all rows of the table must adhere. But in a NoSQL collection, there is no single uniform schema; each document has its own schema and it changes each time a key-value pair is added to or removed from the document. NoSQL collections are said to have **dynamic schemas**.

The program in *Listing 10-4* loads data into the `authors` collection of our `library` database. It reads the data from the same `Authors.dat` file that we used in *Chapter 5, Relational Databases*.

```java
18  public class LoadAuthors {
19      private static final File DATA = new File("data/Authors.dat");
20
21      public static void main(String[] args) {
22          MongoClient client = new MongoClient("localhost", 27017);
23          MongoDatabase library = client.getDatabase("library");
24          MongoCollection authors = library.getCollection("authors");
25
26          authors.drop();
27          library.createCollection("authors");
28          load(authors);
29      }
30
31      public static void load(MongoCollection collection) {...23 lines }
54
55      public static void addDoc(String _id, String lname, String fname, int yob,
56              MongoCollection collection) {...8 lines }
64  }
```

```
 1. AhoA, Aho, Alfred V., 1941
 2. CormenTH, Cormen, Thomas H., 1956
 3. DasguptaS, Dasgupta, Sanjoy, 0
 4. GerstingJ, Gersting, Judith, 0
 5. GoldstineHH, Goldstine, Herman H., 1913
 6. HardyGH, Hardy, Godfrey H., 1877
 7. HopcroftJE, Hopcroft, John E., 1939
 8. HubbardJR, Hubbard, John R., 0
 9. HurayA, Huray, Anita, 0
10. LeisersonCE, Leiserson, Charles E., 1953
11. PapadimitriouC, Papadimitriou, Christos, 0
12. PinsonLJ, Pinson, Lewis J., 0
13. RajaramanA, Rajaraman, Anand, 0
14. RivestRL, Rivest, Ronald L., 1947
```

Listing 10-4. *Java program to load file data into the authors collection*

Lines 22-24 instantiate `MongoClient`, `MongoDatabase`, and `MongoCollection` objects, with the authors object representing the authors collection that we defined in *Figure 10-11*. At lines 26-27, we drop and then re-create the authors collection. Then all the records in the data file are inserted into that collection by the `load(collection)` method called at line 28.

NoSQL Databases

The output shown in Listing 10-4 is generated by the `load(collection)` method, which prints each document (database object) after it is loaded. The code for the load(collection) method is shown in Listing 10-5.

```java
     PrintMongDB.java      InsertMongoDB.java      LoadAuthors.java
31       public static void load(MongoCollection collection) {
32           try {
33               Scanner fileScanner = new Scanner(DATA);
34               int n = 0;
35               while (fileScanner.hasNext()) {
36                   String line = fileScanner.nextLine();
37                   Scanner lineScanner = new Scanner(line).useDelimiter("/");
38                   String _id = lineScanner.next();
39                   String lname = lineScanner.next();
40                   String fname = lineScanner.next();
41                   int yob = lineScanner.nextInt();
42                   lineScanner.close();
43
44                   addDoc(_id, lname, fname, yob, collection);
45                   System.out.printf("%4d. %s, %s, %s, %d%n",
46                           ++n, _id, lname, fname, yob);
47               }
48               System.out.printf("%d docs inserted in authors collection.%n", n);
49               fileScanner.close();
50           } catch (IOException e) {
51               System.err.println(e);
52           }
53       }
```

Listing 10-5. The load() method for the LoadAuthors program

As in *Chapter 5, Relational Databases*, the data is read from the file using `Scanner` objects. The `lineScanner` reads each field separately, as three `String` objects and an `int` value, and then passes them to a separate `addDoc()` method at line 44.

The code for the `addDoc()` method is shown in *Listing 10-6*:

```java
     PrintMongDB.java      InsertMongoDB.java      LoadAuthors.java
55       public static void addDoc(String _id, String lname, String fname, int yob,
56               MongoCollection collection) {
57           Document doc = new Document();
58           doc.put("_id", _id);
59           doc.put("lname", lname);
60           doc.put("fname", fname);
61           doc.put("yob", yob);
62           collection.insertOne(doc);
63       }
```

Listing 10-6. The addDoc() method for the LoadAuthors program

```
> show dbs
admin     0.000GB
friends   0.000GB
library   0.000GB
local     0.000GB
people    0.000GB
> use library
switched to db library
> show collections
authors
books
publishers
> db.authors.find({yob:{$gt:1935}}).sort({yob:-1})
{ "_id" : "SteinCL", "lname" : "Stein", "fname" : "Clifford S.", "yob" : 1965 }
{ "_id" : "CormenTH", "lname" : "Cormen", "fname" : "Thomas H.", "yob" : 1956 }
{ "_id" : "LeisersonCE", "lname" : "Leiserson", "fname" : "Charles E.", "yob" : 1953 }
{ "_id" : "RivestRL", "lname" : "Rivest", "fname" : "Ronald L.", "yob" : 1947 }
{ "_id" : "UllmanJD", "lname" : "Ullman", "fname" : "Jeffrey D.", "yob" : 1942 }
{ "_id" : "AhoA", "lname" : "Aho", "fname" : "Alfred V.", "yob" : 1941 }
{ "_id" : "WienerR", "lname" : "Wiener", "fname" : "Richard", "yob" : 1941 }
{ "_id" : "HopcroftJE", "lname" : "Hopcroft", "fname" : "John E.", "yob" : 1939 }
>
```

Figure 10-16. *MongoDB find() method on authors collection*

The shell capture in *Figure 10-16* confirms that the data was loaded into the `authors` collection. The call `find({yob:{$gt:1935}})` returns all the documents whose `yob` field is greater than 1935.

The `addDoc()` method instantiates a `Document` object at line 57 and then uses its `put()` method for each of the four fields to initialize its corresponding field. Then the statement at line 62 inserts that document into the collection represented by the `collection` parameter.

We can run similar Java programs to load the `publishers` and `books` collections from the `Publishers.dat` and `Books.dat` files, respectively.

NoSQL Databases

When we built the `Library` database as an Rdb in *Chapter 5, Relational Databases*, we also loaded the `AuthorsBooks.dat` file into a separate link table to implement the two foreign keys specified in *Figure 5-2*. But NoSQL databases, like MongoDB do not support foreign keys. The alternative is to include an author or authors field in each `books` document, as shown in *Figure 10-17*. It shows how to add two authors to the book document *Data Structures with Java* that has ISBN 0130933740.

```java
public class AddAuthorsToBooks {
    private static final File AUTHORS_BOOKS = new File("data/AuthorsBooks.dat");

    public static void main(String[] args) {
        MongoClient client = new MongoClient("localhost", 27017);
        MongoDatabase library = client.getDatabase("library");
        MongoCollection books = library.getCollection("books");

        try {
            Scanner scanner = new Scanner(AUTHORS_BOOKS);
            int n = 0;
            while (scanner.hasNext()) {
                String line = scanner.nextLine();
                Scanner lineScanner = new Scanner(line).useDelimiter("/");
                String author_id = lineScanner.next();
                String book_id = lineScanner.next();
                lineScanner.close();

                Document doc = new Document("author_id", author_id);
                books.updateOne(
                        eq("_id", book_id),
                        Updates.addToSet("author", doc));
            }
            scanner.close();
        } catch (IOException e) {
            System.err.println(e);
        }
    }
}
```

Listing 10-7. Adding authors to books documents

First, we use the `find()` method to view the current state of the book document. Note that, at that point, the document has seven fields. Next, we use the `updateOne()` method to add a field with key name `"authors"`. The specified value for that key is the array `["JubbardJR", "HurayA"]`. Finally, we repeat the `find()` query and see that the document now has eight fields, the last one being the added `authors` field.

> The apostrophe (') can be used instead of the quote symbol (") in MongoDB statements, even though the latter is always used for the output from the shell. We prefer the former when typing commands simply because it doesn't require using the *Shift* key.

Alternatively, we could have included code (not shown here) in the Java `LoadBooks` program to read and process the information from the `AuthorsBooks.dat` file.

A Java program that adds all the authors from the `AuthorsBooks.dat` file into the books collection is show in *Listing 10-7*:

```
[> db.books.find({'_id':'0130933740'}).pretty()
{
        "_id" : "0130933740",
        "title" : "Data Structures with Java",
        "edition" : 1,
        "publisher" : "PH",
        "year" : 2004,
        "cover" : "HARD",
        "pages" : 613
}
[> db.books.updateOne({'_id':'0130933740' }, {$set: {'authors':['JubbardJR','HurayA']}})
{ "acknowledged" : true, "matchedCount" : 1, "modifiedCount" : 1 }
[> db.books.find({'_id':'0130933740'}).pretty()
{
        "_id" : "0130933740",
        "title" : "Data Structures with Java",
        "edition" : 1,
        "publisher" : "PH",
        "year" : 2004,
        "cover" : "HARD",
        "pages" : 613,
        "authors" : [
                "JubbardJR",
                "HurayA"
        ]
}
[>
```

Figure 10-17. Using the update() method to add a field to the books collection

Like the previous Java programs in this chapter, this uses two `Scanner` objects to read the data from the specified file. The `while` loop at lines 29-40 reads one line at a time, extracting the `author_id` and `book_id` strings.

The critical code is at lines 36-39. A `Document` object encapsulates the `author_id` at line 36. Then the `updateOne()` method at line 37 adds that `doc` object to the `author` set that belongs to the `books` document identified by `book_id`. If that set does not exist, this method will create it first and then add the `doc` to it.

NoSQL Databases

The Mongo shell session shown in *Figure 10-18* confirms the success of the program in *Listing 10-7*:

```
> db.books.find( {"_id" : "013600637X"} ).pretty()
{
        "_id" : "013600637X",
        "title" : "A First Course in Database Systems",
        "edition" : 3,
        "publisher" : "PH",
        "year" : 2008,
        "cover" : "HARD",
        "pages" : 565
}
> db.books.find( {"_id" : "013600637X"} ).pretty()
{
        "_id" : "013600637X",
        "title" : "A First Course in Database Systems",
        "edition" : 3,
        "publisher" : "PH",
        "year" : 2008,
        "cover" : "HARD",
        "pages" : 565,
        "author" : [
                {
                        "author_id" : "UllmanJD"
                },
                {
                        "author_id" : "WidomJ"
                }
        ]
}
>
```

Figure 10-18. Adding authors to the books collection

The same query, asking for the `books` document that has `_id 013600637X`, is executed twice, first before and then after the execution of that Java program. The second response shows that the authors `UllmanJD` and `WidomJ` have been added to the document's `author` array.

The MongoDB extension for geospatial databases

MongoDB supports the **GeoJSON object types** `Point`, `LineString`, `Polygon`, `MiltiPoint`, `MultiLineString`, `MultiPolygon`, and `GeometryCollection`. These are used in two-dimensional geometry and geographic surface-of-the-earth data.

Mongo provides a nice tutorial on geospatial databases with an application on restaurant locations in New York City here: `https://docs.mongodb.com/manual/tutorial/geospatial-tutorial/`.

A GeoJSON object has the form:

```
<field>: { type: <GeoJSON-type>, coordinates: [longitude, latitude]}
```

Here, `<GeoJSON-type>` is one of the seven types listed previously, and `longitude` and `latitude` are decimal numbers, with range -180 < `longitude` < 180 and -90 < `latitude` < 90. For example, the following is the GeoJSON object for Westminster Abbey in London:

```
"location": {"type": "Point", "coordinates": [-0.1275, 51.4994]}
```

Notice that GeoJSON lists longitude before latitude, as with (x, y) coordinates. This is the reverse of the geo URI scheme that lists latitude first, as shown in *Figure 10-19*:

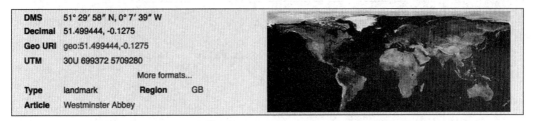

Figure 10-19. The Geo URI for Westminster Abbey: latitude before longitude

The code in *Figure 10-20* illustrates how we can develop a MongoDB collection of places represented by GeoJSON documents:

```
> db.places.insert({
... name:'Greenwich Royal Observatory',
... location:{type:'Point', coordinates:[0.0, 51.4768]},
... category:'Astronomical Observatory'
... })
WriteResult({ "nInserted" : 1 })
> db.places.find({},{_id:0})
{ "name" : "Greenwich Royal Observatory", "location" : { "type" : "Point", "coordinates" : [ 0, 51.4768 ] }, "category" : "Astronomical Observatory" }
>
```

Figure 10-20. Inserting a GeoJSON document into a collection of places

Indexing in MongoDB

Recall that an index on a database field is a tree structure that greatly increases the efficiency of queries on that field. In *Chapter 5, Relational Databases*, we described the B-tree data structure (see *Figure 5-18*) that is the common mechanism for implementing database indexes.

Like relational database systems, Mongo also supports indexing. As an example, suppose that our `books` collection contains 1,000,000 documents, one for each book. Further, suppose that this query is executed:

```
db.books.find({year:{"$gte":1924,"$lt":1930}})
```

This would list all books published from 1924 to 1930. If the year filed is indexed, the response would be instantaneous. Otherwise, every one of the 1,000,000 documents would have to be examined.

The required `_id` field in each collection is indexed automatically. This is a unique index, which means that it prevents the insertion of any document with the same `_id` value as one already in the collection.

To index any other field, use the `db.collection.createIndex(key)` method, as shown in *Figure 10-19*. The value `1` indicates that the index is in an ascending order of the specified field values.

```
> db.books.createIndex({'year':1})
{
        "createdCollectionAutomatically" : false,
        "numIndexesBefore" : 1,
        "numIndexesAfter" : 2,
        "ok" : 1
}
>
```

Figure 10-21. Creating an index on the books.year field

As in relational databases, indexes take up a lot of space, and they can slow the process of insertions and deletions. So, it's probably not a good idea to index every field. The best strategy is to create indexes on just the most commonly searched fields. For example, in our `library` database, we would probably want to create indexes on the `books.year` field, the `books.author.author_id` field, the `books.title` field, and the `publishers.name` field.

MongoDB also supports **compound indexes**. The general syntax is:

```
db.collection.createIndex({<field1>: <type>, <field2>: <type2> ... })
```

For example, `db.books.createIndex({year:1, title:1})` would create a two-dimensional compound index, indexing first on the `year` field, and second on the `title` field within each year. That index would facilitate frequent queries like this:

```
db.books.find({}, {year:1, title:1, publisher:1}).sort({year:1})
```

We can also index geospatial database collections. MongoDB supports two specialized geospatial indexes: one for planar two-dimensional geometric data, and one for spherical geometric data. The former is useful in graphics applications, while the latter applies to geographic locations on the surface of the earth. To see why these two contexts are different, recall that the sum of the angles of a plane triangle is always 180°; but in a spherical triangle, all three angles could be right angles, making the sum 270°. Think of a spherical triangle with its base on the equator and its two sides on meridians down from the North Pole.

Why NoSQL and why MongoDB?

In the past decade, dataset sizes, especially for web-based enterprises, have been growing incredibly fast. Storage demands are now in the terabyte range. As this demand increases, developers have to choose between expanding their machine size (scaling up) or distributing their data among many independent machines (scaling out). A growing database is easier to manage, scaling up, but that option is more expensive and eventually limited in size. Scaling out is clearly a better option, but standard Rdbs do not distribute easily or cheaply.

MongoDB is a document-based system that easily scales out. It automatically balances its databases across a cluster, redistributing documents transparently, making it easy to add machines when needed.

Other NoSQL database systems

As mentioned earlier, MongoDB is currently the top ranked NoSQL database system. (see http://www.kdnuggets.com/2016/06/top-nosql-database-engines.html) It is followed by Apache Cassandra, Redis, Apache HBase, and Neoj4.

MongoDB uses the **document data model**: a database is a set of collections, each of which is a set of documents, each of which is a set of key-value pairs. Each document is stored as a BSON file.

Cassandra and HBase use the **column data model**: each data element is a triple: a key, its value, and a timestamp. Cassandra has its own query language, called CQL, that looks like SQL.

Redis uses the **key-value data model**: a database is a set of dictionaries, each of which is a set of key-value records, where the value is a sequence of fields.

Neoj4 uses the **graph data model**: a database is a graph whose nodes contain the data. It supports the Cypher query language. Other graph DBS that support Java include GraphBase and OrientDB.

These all have Java APIs, so you can write essentially the same Java access programs that we have used here for our MongoDB database.

Summary

This chapter introduces NoSQL databases and the MongoDB database system. It discusses the differences between relational SQL databases and nonrelational NoSQL databases, their structure and their uses. Our Library database from *Chapter 5, Relational Databases*, is rebuilt as a MongoDB database, and Java application programs are run against it. Finally, we briefly consider geospatial databases and indexing in MongoDB.

11
Big Data Analysis with Java

"In pioneer days they used oxen for heavy pulling, and when one ox couldn't budge a log, they didn't try to grow a larger ox. We shouldn't be trying for bigger computers, but for more systems of computers."

– Grace Hopper (1906-1992)

The term big data generally refers to algorithms for the storage, retrieval, and analysis of massive datasets that are too large to be managed by a single file server. Commercially, these algorithms were pioneered by Google. Two of their early benchmark algorithms, PageRank and MapReduce, are among those considered in this chapter.

> The word "googol" was coined by the nine-year-old nephew of the American mathematician Edward Kasner in the 1930s. The word was meant to stand for 10^{100}. At that time, it had been estimated that the number of particles in the universe was about 10^{80}. Kasner later coined the additional word "googolplex" to stand for 10^{google}. That would be written as 1 followed by 10^{100} zeros. Google's new headquarters in Mountain View, California is called the Googleplex.

Scaling, data striping, and sharding

Relational database systems (Rdbs) are not very good at managing very large databases. As we saw in *Chapter 10, NoSQL Databases*, that was one major reason why NoSQL database systems were developed.

There are two approaches to managing increasingly large datasets: vertical scaling and horizontal scaling. **Vertical scaling** refers to the strategy of increasing the capacity of a single server by upgrading to more powerful CPUs, more main memory, and more storage space. **Horizontal scaling** refers to the redistribution of the dataset by increasing the number of servers in the system. Vertical scaling has the advantage that it does not require any significant retooling of existing software; its main disadvantage is that it is more tightly limited than horizontal scaling. The main problem with horizontal scaling is that it does require adjustments in the software. But as we have seen, frameworks like MapReduce have made many of those adjustments quite manageable.

Data striping refers to the distribution of relatively small chunks of data across several storage devices, such as hard disks on a single machine. It is a process that has long been used by relational database systems to facilitate faster record access and to provide data redundancy.

When horizontal scaling is used to accommodate very large datasets, the data is systematically distributed over the cluster. MongoDB uses a technique called **sharding** to do this. It partitions a collection of documents into subsets, called **shards**.

Sharding can be done in two ways in MongoDB: hashed sharding and range sharding. This dichotomy is analogous to the choice in Java of a `HashMap` or a `TreeMap` for implementing the `Map` interface. The preference usually depends upon whether you will want to do range queries on your collection key. For example, the key for our `library.books` collection is `isbn`. We're not likely to have a query such as "Find all books with ISBN between 1107015000 and 1228412000". So, hash sharding would be better for that collection. On the other hand, if we key on the `year` field, a range query might be likely, so in that case, range sharding might be better.

Google's PageRank algorithm

Within a few years of the birth of the web in 1990, there were over a dozen search engines that users could use to search for information. Shortly after it was introduced in 1995, AltaVista became the most popular among them. These search engines would categorize web pages according to the topics that the pages themselves specified.

But the problem with these early search engines was that unscrupulous web page writers used deceptive techniques to attract traffic to their pages. For example, a local rug-cleaning service might list "pizza" as a topic in their web page header, just to attract people looking to order a pizza for dinner. These and other tricks rendered early search engines nearly useless.

Chapter 11

To overcome the problem, various page ranking systems were attempted. The objective was to rank a page based upon its popularity among users who really did want to view its contents. One way to estimate that is to count how many other pages have a link to that page. For example, there might be 100,000 links to https://en.wikipedia.org/wiki/Renaissance, but only 100 to https://en.wikipedia.org/wiki/Ernest_Renan, so the former would be given a much higher rank than the latter.

But simply counting the links to a page will not work either. For example, the rug-cleaning service could simply create 100 bogus web pages, each containing a link to the page they want users to view.

In 1996, Larry Page and Sergey Brin, while students at Stanford University, invented their PageRank algorithm. It simulates the web itself, represented by a very large directed graph, in which each web page is represented by a node in the graph, and each page link is represented by a directed edge in the graph.

[Their work on search engines led to the founding of Google in 1998.]

The directed graph shown in *Figure 11-1* could represent a very small network with the same properties:

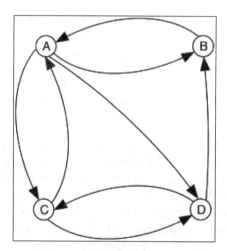

Figure 11-1. A directed graph

This has four nodes, representing four web pages, **A**, **B**, **C**, and **D**. The arrows connecting them represent page links. So, for example, page **A** has a link to each of the other three pages, but page **B** has a link only to **A**.

To analyze this tiny network, we first identify its **transition matrix**, M:

$$M = \begin{bmatrix} 0 & 1 & \tfrac{1}{2} & 0 \\ \tfrac{1}{3} & 0 & 0 & \tfrac{1}{2} \\ \tfrac{1}{3} & 0 & 0 & \tfrac{1}{2} \\ \tfrac{1}{3} & 0 & \tfrac{1}{2} & 0 \end{bmatrix}$$

This square has 16 entries, m_{ij}, for $1 \le i \le 4$ and $1 \le j \le 4$. If we assume that a web crawler always picks a link at random to move from one page to another, then m_{ij} equals the probability that it will move to node i from node j, (numbering the nodes A, B, C, and D as 1, 2, 3, and 4). So $m_{12} = 1$ means that if it's at node B, there's a 100% chance that it will move next to A. Similarly, $m_{13} = m_{43} = \tfrac{1}{2}$ means that if it's at node C, there's a 50% chance of it moving to A and a 50% chance of it moving to D.

Suppose a web crawler picks one of those four pages at random, and then moves to another page, once a minute, picking each link at random. After several hours, what percentage of the time will it have spent at each of the four pages?

Here is a similar question. Suppose there are 1,000 web crawlers who obey that transition matrix as we've just described, and that 250 of them start at each of the four pages. After several hours, how many will be on each of the four pages?

This process is called a **Markov chain**. It is a mathematical model that has many applications in physics, chemistry, computer science, queueing theory, economics, and even finance.

The diagram in *Figure 11-1* is called the state diagram for the process, and the nodes of the graph are called the states of the process. Once the state diagram is given, the meaning of the nodes (web pages, in this case) becomes irrelevant. Only the structure of the diagram defines the transition matrix M, and from that we can answer the question. A more general Markov chain would also specify transition probabilities between the nodes, instead of assuming that all transition choices are made at random. In that case, those transition probabilities become the non-zero entries of the M.

A Markov chain is called **irreducible** if it is possible to get to any state from any other state. With a little checking, you can see that the Markov chain defined by the graph in *Figure 11-1* is irreducible. This is important because the mathematical theory of Markov chains tells us that if the chain is irreducible, then we can compute the answer to the preceding question using the transition matrix.

What we want is the steady state solution; that is, a distribution of crawlers that doesn't change. The crawlers themselves will change, but the number at each node will remain the same.

To calculate the steady state solution mathematically, we first have to realize how to apply the transition matrix M. The fact is that if $\mathbf{x} = (x_1, x_2, x_3, x_4)$ is the distribution of crawlers at one minute, and the next minute the distribution is $\mathbf{y} = (y_1, y_2, y_3, y_4)$, then $\mathbf{y} = M\mathbf{x}$, using matrix multiplication.

For example, if 30% of the crawlers are at node A and 70% are at node B at one minute, then in the next minute, 70% will be at A, 10% at B, 10% at C, and 10% at D:

$$M\mathbf{x} = \begin{bmatrix} 0 & 1 & 1/2 & 0 \\ 1/3 & 0 & 0 & 1/2 \\ 1/3 & 0 & 0 & 1/2 \\ 1/3 & 0 & 1/2 & 0 \end{bmatrix} \begin{bmatrix} 0.3 \\ 0.7 \\ 0.0 \\ 0.0 \end{bmatrix} = \begin{bmatrix} (0)(0.3)+(1)(0.7)+(1/2)(0.0)+(0)(0.0) \\ (1/3)(0.3)+(0)(0.7)+(0)(0.0)+(1/2)(0.0) \\ (1/3)(0.3)+(0)(0.7)+(0)(0.0)+(1/2)(0.0) \\ (1/3)(0.3)+(0)(0.7)+(1/2)(0.0)+(0)(0.0) \end{bmatrix} = \begin{bmatrix} 0.7 \\ 0.1 \\ 0.1 \\ 0.1 \end{bmatrix}$$

These are just conditional probability calculations (note that vectors are expressed as columns here).

So now, if $\mathbf{x}$ is the steady state solution for the Markov chain, then $M\mathbf{x} = \mathbf{x}$. This vector equation gives us four scalar equations in four unknowns:

$$M\mathbf{x} = \begin{bmatrix} 0 & 1 & 1/2 & 0 \\ 1/3 & 0 & 0 & 1/2 \\ 1/3 & 0 & 0 & 1/2 \\ 1/3 & 0 & 1/2 & 0 \end{bmatrix} \begin{bmatrix} x_1 \\ x_2 \\ x_3 \\ x_4 \end{bmatrix} = \begin{bmatrix} x_1 \\ x_2 \\ x_3 \\ x_4 \end{bmatrix}$$

One of these equations is redundant (linearly dependent). But we also know that $x_1 + x_2 + x_3 + x_4 = 1$, since **x** is a probability vector. So, we're back to four equations in four unknowns. The solution is:

$$\mathbf{x} = \begin{bmatrix} 1/3 \\ 2/9 \\ 2/9 \\ 2/9 \end{bmatrix}$$

The point of that example is to show that we can compute the steady state solution to a static Markov chain by solving an $n \times n$ matrix equation, where n is the number of states. By static here, we mean that the transition probabilities m_{ij} do not change. Of course, that does not mean that we can mathematically compute the web. In the first place, $n > 30,000,000,000,000$ nodes! And in the second place, the web is certainly not static. Nevertheless, this analysis does give some insight about the web; and it clearly influenced the thinking of Larry Page and Sergey Brin when they invented the PageRank algorithm.

Another important distinction between the web and the previous example is that the transition matrix for the web is very sparse. In other words, nearly all the transition probabilities are zero.

A sparse matrix is usually represented as a list of key-value pairs, where the key identifies a node, and its value is the list of nodes that can be reached from that node in one step. For example, the transition matrix M for the previous example would be represented as shown in *Table 11-1*.

Key	Value
A	B, C, D
B	A
C	A, D
D	B, C

Table 11-1. Adjacency list

As we have seen, this type of data structure is very amenable to the MapReduce framework, which we examine in the next section of this chapter.

Recall that the purpose of the PageRank algorithm is to rank the web pages according to some criteria that would resemble their importance, or at least their frequency of access. The original simple (pre-PageRank) idea was to count the number of links to each page and use something proportional to that count for the rank. Following that line of thought, we can imagine that, if $\mathbf{x} = (x_1, x_2, ..., x_n)^T$ is the page rank for the web (that is, if x_j is the relative rank of page j and $\Sigma x_j = 1$), then $M\mathbf{x} = \mathbf{x}$, at least approximately. Another way to put that is that repeated applications of M to $\mathbf{x}$ should nudge $\mathbf{x}$ closer and closer to that (unattainable) steady state.

That brings us (finally) to the PageRank formula:

$$x' = f(\mathbf{x}) = (1-\varepsilon)M\mathbf{x} + \varepsilon\, z/n$$

where ε is a very small positive constant, $\mathbf{z}$ is the vector of all 1s, and n is the number of nodes. The vector expression on the right defines the transformation function f which replaces a page rank estimate $\mathbf{x}$ with an improved page rank estimate. Repeated applications of this function gradually converge to the unknown steady state.

Note that in the formula, f is a function of more than just $\mathbf{x}$. There are really four inputs: $\mathbf{x}$, M, ε, and n. Of course, $\mathbf{x}$ is being updated, so it changes with each iteration. But M, ε, and n change too. M is the transition matrix, n is the number of nodes, and ε is a coefficient that determines how much influence the $\mathbf{z}/n$ vector has. For example, if we set ε to 0.00005, then the formula becomes:

$$x' = 0.99995\, M\mathbf{x} + 0.00005\, z/n$$

Google's MapReduce framework

How do you quickly sort a list of a billion elements? Or multiply two matrices, each with a million rows and a million columns?

In implementing their PageRank algorithm, Google quickly discovered the need for a systematic framework for processing massive datasets. That can be done only by distributing the data and the processing over many storage units and processors. Implementing a single algorithm, such as PageRank in that environment is difficult, and maintaining the implementation as the dataset grows is even more challenging.

The answer is to separate the software into two levels: a framework that manages the big data access and parallel processing at a lower level, and a couple of user-written methods at an upper-level. The independent user who writes the two methods need not be concerned with the details of the big data management at the lower level.

Specifically, the data flows through a sequence of stages:

1. The input stage divides the input into chunks, usually 64 MB or 128 MB.
2. The mapping stage applies a user-defined `map()` function that generates from one key-value pair a larger collection of key-value pairs of a different type.
3. The partition/grouping stage applies hash sharding to those keys to group them.
4. The reduction stage applies a user-defined `reduce()` function to apply some specific algorithm to the data in the value of each key-value pair.
5. The output stage writes the output from the `reduce()` method.

The user's choice of `map()` and `reduce()` methods determines the outcome of the entire process; hence the name **MapReduce**.

This idea is a variation on the old algorithmic paradigm called **divide and conquer**. Think of the proto-typical mergesort, where an array is sorted by repeatedly dividing it into two halves until the pieces have only one element, and then they are systematically pairwise merged back together.

MapReduce is actually a meta-algorithm—a framework, within which specific algorithms can be implemented through its `map()` and `reduce()` methods. Extremely powerful, it has been used to sort a petabyte of data in only a few hours. Recall that a petabyte is $1000^5 = 10^{15}$ bytes, which is a thousand terabytes or a million gigabytes.

Some examples of MapReduce applications

Here are a few examples of big data problems that can be solved with the MapReduce framework:

1. Given a repository of text files, find the frequency of each word. This is called the **WordCount** problem.
2. Given a repository of text files, find the number of words of each word length.
3. Given two matrices in sparse matrix format, compute their product.
4. Factor a matrix given in sparse matrix format.
5. Given a symmetric graph whose nodes represent people and edges represent friendship, compile a list of common friends.
6. Given a symmetric graph whose nodes represent people and edges represent friendship, compute the average number of friends by age.
7. Given a repository of weather records, find the annual global minima and maxima by year.
8. Sort a large list. Note that in most implementations of the MapReduce framework, this problem is trivial, because the framework automatically sorts the output from the `map()` function.
9. Reverse a graph.
10. Find a **minimal spanning tree** (**MST**) of a given weighted graph.
11. Join two large relational database tables.

Examples *9* and *10* apply to graph structures (nodes and edges). For very large graphs, more efficient methods have been developed recently; for example, the Apache Hama framework created by Edward Yoon.

The WordCount example

In this section, we present the MapReduce solution to the WordCount problem, sometimes called the Hello World example for MapReduce.

The diagram in *Figure 11-2* shows the data flow for the WordCount program. On the left are two of the 80 files that are read into the program:

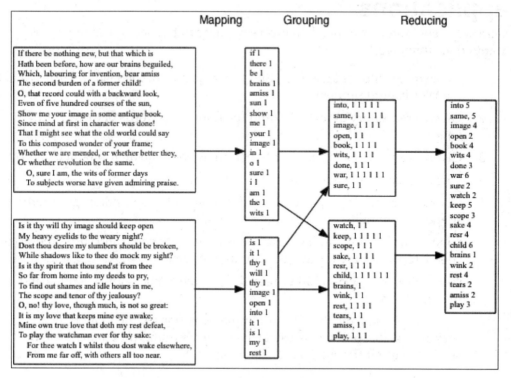

Figure 11-2. Data flow for the WordCount program

During the mapping stage, each word, followed by the number 1, is copied into a temporary file, one pair per line. Notice that many words are duplicated many times. For example, `image` appears five times among the 80 files (including both files shown), so the string `image 1` will appear four times in the temporary file. Each of the input files has about 110 words, so over 8,000 word-number pairs will be written to the temporary file.

Note that this figure shows only a very small part of the data involved. The output from the mapping stage includes every word that is input, as many times that it appears. And the output from the grouping stage includes every one of those words, but without duplication.

The grouping process reads all the words from the temporary file into a key-value hash table, where the key is the word, and the value is a string of 1s, one for each occurrence of that word in the temporary file. Notice that these 1s written to the temporary file are not used. They are included simply because the MapReduce framework in general expects the `map()` function to generate key-value pairs.

The reducing stage transcribed the contents of the hash table to an output file, replacing each string of 1s with the number of them. For example, the key-value pair `("book", "1 1 1 1")` is written as `book 4` in the output file.

Keep in mind that this is a toy example of the MapReduce process. The input consists of 80 text files containing about 9073 words. So, the temporary file has 9073 lines, with one word per line. Only 2149 of those words are distinct, so the hash table has 2149 entries and the output file has 2149 lines, with one word per line.

The program in *Listing 11-1* implements the WordCount solution that is consistent with the MapReduce framework.

```java
public class Example1 {
    public static void main(String[] args) {
        try {
            File tempFile = new File("data/Temp.dat");
            map("data/sonnets/", 80, tempFile);

            Map<String,StringBuilder> hashTable = new HashMap(2500);
            combine(tempFile, hashTable);

            File outFile = new File("data/Output.dat");
            reduce(hashTable, outFile);
        } catch (IOException e) {
            System.err.println(e);
        }
    }

    public static void map(String src, int n, File temp)
            throws IOException {...8 lines }

    public static void combine(File temp, Map<String,StringBuilder> table)
            throws IOException {...13 lines }

    public static void reduce(Map<String,StringBuilder> table, File out)
            throws IOException {...9 lines }

    /* Writes the pair (word, 1) for each word in the specified file.
    */
    public static void map(String filename, PrintWriter writer)
            throws IOException {...9 lines }

    /* Counts the 1s in the value argument and then writes (key, count).
    */
    public static void reduce(String key, String value, PrintWriter writer)
            throws IOException {...4 lines }

    private static void sort(File file) throws IOException {...14 lines }
}
```

Listing 11-1. WordCount program

In addition to the 80 text files in the directory data/sonnets/, the program uses two other files: data/Temp.dat and data/Output.dat, declared at lines 21 and 27.

It also uses a hash table, defined at line 24. The main() method performs three tasks: mapping, combining, and reducing, executed by the method calls at lines 22, 25, and 28. Those implementations are shown in *Listing 11-2*:

```java
34      public static void map(String src, int n, File temp)
35              throws IOException {
36          PrintWriter writer = new PrintWriter(temp);
37          for (int i = 0; i < n; i++) {
38              String filename = String.format("%sSonnet%03d.txt", src, i+1);
39              map(filename, writer);
40          }
41          writer.close();
42      }
43
44      public static void combine(File temp, Map<String,StringBuilder> table)
45              throws IOException {
46          Scanner scanner = new Scanner(temp);
47          while (scanner.hasNext()) {
48              String word = scanner.next();
49              StringBuilder value = table.get(word);
50              if (value == null) {
51                  value = new StringBuilder("");
52              }
53              table.put(word, value.append(" 1"));
54              scanner.nextLine();   // scan past the rest of the line (a "1")
55          }
56          scanner.close();
57      }
58
59      public static void reduce(Map<String,StringBuilder> table, File out)
60              throws IOException {
61          PrintWriter writer = new PrintWriter(out);
62          for (Map.Entry<String, StringBuilder> entry : table.entrySet()) {
63              String key = entry.getKey();   // e.g., "speak"
64              String value = entry.getValue().toString();   // e.g., "1 1 1 1 1"
65              reduce(key, value, writer);
66          }
67          writer.close();
68      }
```

Listing 11-2. Methods for the WordCount program

The `map()` method, implemented at lines 34-42, simply applies another `map()` method to each of the 80 files in the `data/sonnets/` directory. Note that the output for this inner `map()` method is specified by the `writer` parameter, which is a `PrintWriter` object. It could be a writer to a file, as it is here, or to a string, or to any other more general `OutputStream` object. The choice is made by the calling method, which is the outer `map()` method in this case.

The output from the outer `map()` method is illustrated in *Figure 11-2*. It consists of a large number of key-value pairs, where the key is one of the words read from one of the 80 input files, such as `amiss`, and the value is the integer 1.

The `combine()` method, implemented at lines 44-57, reads all the lines from the specified `temp` file, expecting each to be a word followed by the integer 1. It loads each of these words into the specified hash table, where the value of each word key is a string of 1s, one for each occurrence of the word found, as illustrated in *Figure 11-2*.

Notice how the code at lines 49-53 works. If the word has already been put into the hash table, then `put()` at line 53 will simply append one more 1 to the sequence of 1s that it already has for the value of that key. But the first time the word is read from the file, the `get()` method at line 49 will return `null`, causing line 51 to execute, resulting in a single 1 being inserted with that word at line 53.

In Java, if a key is already in a `HashMap` (that is, a hash table), then the `put()` method will simply update the value of the existing key-value pair. This prevents duplicate keys from being inserted.

The `reduce()` method, implemented at lines 59-68, carries out the reducing stage of the MapReduce process, as we described previously. For example, it will read a key-value pair such as (`"book"`, `"1 1 1 1"`) from the hash table and then write it to the output file as `book 4`. This is done by an inner `reduce()` method, called at line 65.

The code in *Listing 11-3* shows the inner `map()` and `reduce()` methods, called at lines 39 and 65. These two elementary methods were described previously.

```java
      /* Writes the pair (word, 1) for each word in the specified file.
       */
      public static void map(String filename, PrintWriter writer)
              throws IOException {
          Scanner input = new Scanner(new File(filename));
          input.useDelimiter("[.,:;()?!\"\\s]+");
          while (input.hasNext()) {
              String word = input.next();
              writer.printf("%s 1%n", word.toLowerCase());
          }
          input.close();
      }


      /* Counts the 1s in the value argument and then writes (key, count).
       */
      public static void reduce(String key, String value, PrintWriter writer)
              throws IOException {
          int count = (value.length() + 1)/2;  // e.g. "1 1 1 1 1" => 5
          writer.printf("%s %d%n", key, count);
      }
```

Listing 11-3. The map and reduce methods for the WordCount program

Of course, you could write a simpler program to count the frequencies of words in a directory of files. But one point of this example is to show how to write `map()` and `reduce()` methods that conform to the MapReduce framework.

Scalability

The great benefit of the MapReduce framework is that it is **scalable**. The WordCount program in `Example1.java` was run on 80 files containing fewer than 10,000 words. With little modification, it could be run on 80,000 files with 10,000,000 words. That flexibility in software is called **scalability**.

To manage that thousand-fold increase in input, the hash table might have to be replaced. Even if we had enough memory to load a table that large, the Java processing would probably fail because of the proliferation of objects. Object-oriented programming is certainly the best way to implement an algorithm. But if you want clarity, speed, and flexibility it is not so efficient at handling large datasets.

We don't really need the hash table, which is instantiated at line 24 in *Listing 11-1*. We can implement the same idea by hashing the data into a set of files instead. This is illustrated in *Figure 11-3*.

Replacing the hash table with file chunks would require modifying the code at lines 34-68 (*Listing 11-2*), but not in the `map()` and `reduce()` methods at lines 70-89 (*Listing 11-3*). And those two methods are where the actual number crunching takes place; everything else is just moving things around. It's only the `map()` and `reduce()` methods that actually determine that we are counting words here.

So, this is the main idea of the MapReduce meta-algorithm: provide a framework for processing massive datasets, a framework that allows the independent programmer to *plug in* specialized `map()` and `reduce()` methods that actually implement the required particular algorithm. If that particular algorithm is to count words, then write the `map()` method to extract each individual word from a specified file and write the key-value pair (`word, 1`) to wherever the specified `writer` will put them, and write the `reduce()` method to take a key-value pair such as (`word, 1 1 1 1`) and return the corresponding key-value pair as (`word, 4`) to wherever its specified `writer` will put it. These two methods are completely localized—they simply operate on key-value pairs. And, they are completely independent of the size of the dataset.

The diagram in *Figure 11-3* illustrates the general flow of data through an application of the MapReduce framework:

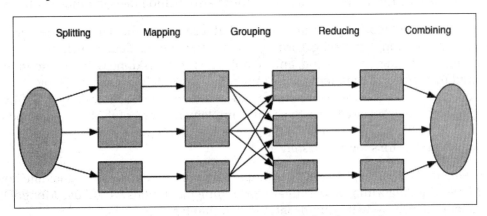

Figure 11-3. Data flow in the general MapReduce framework

The original dataset could be in various forms and locations: a few files in a local directory, a large collection of files distributed over several nodes on the same cluster, a database on a database system (relational or NoSQL), or data sources available on the World Wide Web. The MapReduce controller then carries out these five tasks:

1. Split the data into smaller datasets, each of which can be easily accessed on a single machine.
2. Simultaneously (that is, in parallel), run a copy of the user-supplied `map()` method, one on each dataset, producing a set of key-value pairs in a temporary file on that local machine.
3. Redistribute the datasets among the machines, so that all instances of each key are in the same dataset. This is typically done by hashing the keys.
4. Simultaneously (in parallel), run a copy of the user-supplied `reduce()` method, one on each of the temporary files, producing one output file on each machine.
5. Combine the output files into a single result. If the `reduce()` method also sorts its output, then this last step could also include merging those outputs.

The genius of the MapReduce framework is that it separates the data management (moving, partitioning, grouping, sorting, and so on) from the data crunching (counting, averaging, maximizing, and so on). The former is done with no attention required by the user. The latter is done in parallel, separately in each node, by invoking the two user-supplied methods `map()` and `reduce()`. Essentially, the only obligation of the user is to devise the correct implementations of these two methods that will solve the given problem. In the case of the WordCount problem, those implementations are shown in *Listing 11-3*.

As we mentioned earlier, these code examples are presented mainly to elucidate how the MapReduce algorithm works. Real-world implementations would use MongoDB or Hadoop, which we will examine later in this chapter.

Matrix multiplication with MapReduce

If A is an $m \times p$ matrix and B is an $p \times n$ matrix, then the product of A and B is the $m \times n$ matrix $C = AB$, where the $(i, j)^{th}$ element of C is computed as the inner product of the ith row of A with the jth column of B:

$$c_{ij} = \sum_{k=1}^{p} a_{ik} b_{kj}$$

This is a dot product—simple arithmetic if m, p, and n are small. But not so simple if we're working with big data.

The formula for c_{ij} requires p multiplications and $p - 1$ additions, and there are $m \cdot n$ of these to do. So, that implementation runs in $O(mnp)$ time. That is slow. Furthermore, if A and B are dense matrices (that is, most elements are nonzero), then storage requirements can also be overwhelming. This looks like a job for MapReduce.

For MapReduce, think key-value pairs. We assume that each matrix is stored as a sequence of key-value pairs, one for each non-zero element of the matrix. The key is the subscript pair (i, j), and the value is the $(i, j)^{th}$ element of the matrix. For example, this matrix

$$A = \begin{bmatrix} 0 & 7.23 & 0 & 9.11 & 4.54 \\ 0 & 0 & 6.87 & 0 & 0 \\ 4.09 & 0 & 0 & 0 & 0 \\ 1.54 & 0 & 0 & 0 & 3.36 \end{bmatrix}$$

would be represented by the list shown in *Figure 11-4*. This is sometimes called **sparse matrix format**.

```
MatrixA.dat
(1, 2)   7.23
(1, 4)   9.11
(1, 5)   4.54
(2, 3)   6.87
(3, 1)   4.09
(4, 1)   1.54
(4, 5)   3.36
```

Figure 11-4. Key-value pairs

By list, we don't necessarily mean a Java `List` object. In practice, it could be a file or more general input stream. In the following coding, we will assume that each is a text file, represented by a Java `File` object.

To implement matrix multiplication within the MapReduce framework, we will assume that we are given a file containing the elements of two matrices, organized as shown in *Figure 11-4*. These 12 values are the elements of a 3 × 2 matrix named A followed by a 2 × 3 matrix named B. For example, $a_{12} = 3.21$ and $b_{23} = 1.94$. Notice that these two matrices are full—no zero elements.

Big Data Analysis with Java

The `map()` and `reduce()` methods for multiplying two matrices that are input this way are shown in *Listing 11-4*.

```java
   TwoMatrices   Example2.java
72      /* Reads ("a", i, k, x), representing array element x = a[i,k],
73         and writes key = (i, j) and value x, for j = 1..n;
74         then reads ("b", k, j, y), representing y = b[k,j]
75         and writes key = (i, j) and value y, for i = 1..m.
76      */
77      public static void map(String element, PrintWriter writer)
78              throws IOException {
79          Scanner input = new Scanner(new File(element));
80          String name = input.next();    // "a" or "b"
81          if (name.equals("a")) {
82              int i = input.nextInt();
83              int k = input.nextInt();
84              double x = input.nextDouble();    // x = a[i,k]
85              for (int j = 1; j <= n; j++) {
86                  writer.printf("(%d,%d), %.4f%n", i, j, x);
87              }
88          } else {  // name = "b"
89              int k = input.nextInt();
90              int j = input.nextInt();
91              double y = input.nextDouble();    // y = b[j,k]
92              for (int i = 1; i <= m; i++) {
93                  writer.printf("(%d,%d), %.4f%n", i, j, y);
94              }
95          }
96          input.close();
97      }
98
99      /* For a key (i,j) the value will be:
100             "a[i,1] a[i,2] ... a[i,p] b[1,j] b[2,j] ... b[p,j]".
101            Reduces to a[i,1]*b[1,j] + a[i,2]*b[2,j] + ... + a[i,p]*b[p,j].
102     */
103     public static void reduce(String key, String value, PrintWriter writer)
104             throws IOException {
105         double[] x = new double[p];
106         double[] y = new double[p];
107         Scanner scanner = new Scanner(value);
108         for (int k = 0; k < p; k++) {
109             x[k] = scanner.nextDouble();
110         }
111         for (int k = 0; k < p; k++) {
112             y[k] = scanner.nextDouble();
113         }
114         double sum = 0.0;
115         for (int k = 0; k < p; k++) {
116             sum += x[k]*y[k];
117         }
118         writer.printf("%s %.4f%n", key, sum);
119     }
```

Listing 11-4. *The map and reduce methods for matrix multiplication*

The complete MapReduce program for this application is similar to the WordCount implementation shown in *Listing 11-1*.

As the comment at lines 72-76 indicates, the `map()` method reads the input file, one line at a time. For example, the first line from the file shown in *Figure 11-5* would be:

```
a 1 1 4.26
```

```
TwoMatrices
a 1 1 4.26
a 1 2 3.21
a 2 1 7.08
a 2 2 1.94
a 3 1 5.01
a 3 2 7.25
b 1 1 6.88
b 1 2 7.02
b 1 3 4.23
b 2 1 5.01
b 2 2 6.88
b 2 3 1.94
```

Figure 11-5. Two matrices

The numeric values would be stored as $i = 1$, $k = 1$, and $x = 4.26$ (at lines 82-84). Then the for loop at lines 85-87 would write these three outputs to whatever context is assigned to the `writer` object:

```
(1,1)    4.26
(1,2)    4.26
(1,3)    4.26
```

Note that, for this example, the program has set the global constants $m = 3$, $p = 2$, and $n = 3$ for the dimensions of the matrices.

The `map()` method writes each value three times because each one will be used in three different sums.

The grouping process that follows the `map()` calls will reassemble the data like this:

(1,1) a_{11} a_{12} b_{11} b_{21}

(1,2) a_{11} a_{12} b_{12} b_{22}

(1,3) a_{11} a_{12} b_{13} b_{23}

(2,1) a_{21} a_{22} b_{11} b_{21}

(2,2) a_{21} a_{22} b_{12} b_{22}

Then for each key (i, j), the `reduce()` method computes the inner product of the two vectors that are listed for that key's value:

(1,1) $a_{11}b_{11} + a_{12}b_{21}$

(1,2) $a_{11}b_{12} + a_{12}b_{22}$

(1,3) $a_{11}b_{13} + a_{12}b_{23}$

(2,1) $a_{21}b_{11} + a_{22}b_{21}$

(2,2) $a_{21}b_{12} + a_{22}b_{22}$

Those are the correct values for the elements c_{11}, c_{12}, c_{13}, and so on of the product matrix $C = AB$.

 Note that for this to work correctly, the order in which the `map()` method emits the key-values must be maintained by the grouping process. If it doesn't, then some additional indexing scheme may have to be included so that the `reduce()` method can do the $a_{ik} b_{kj}$ pairing correctly.

MapReduce in MongoDB

MongoDB implements the MapReduce framework with its `mapReduce()` command. An example is shown in *Figure 11-6*.

```
> var map1 = function() { emit(this.publisher, 1); };
> var reduce1 = function(pubId, numBooks) { return Array.sum(numBooks); };
> db.books.mapReduce(map1, reduce1, {out: "map_reduce_example"}).find()
{ "_id" : "A-V", "value" : 1 }
{ "_id" : "A-W", "value" : 3 }
{ "_id" : "BACH", "value" : 1 }
{ "_id" : "CAMB", "value" : 2 }
{ "_id" : "EDIS", "value" : 1 }
{ "_id" : "MHE", "value" : 4 }
{ "_id" : "OXF", "value" : 1 }
{ "_id" : "PH", "value" : 3 }
{ "_id" : "PUP", "value" : 1 }
{ "_id" : "TEUB", "value" : 1 }
{ "_id" : "WHF", "value" : 1 }
>
>
```

Figure 11-6. Running MapReduce in MongoDB

The first two statements define the JavaScript functions `map1()` and `reduce1()`. The third statement runs MapReduce on our `library.books` collection (see *Chapter 10, NoSQL Databases*), applying those two functions, and naming the resulting collection `"map_reduce_example"`. Appending the `find()` command causes the output to be displayed.

The `map1()` function emits the key-value pair `(p, 1)`, where p is the `books.publisher` field. So this will generate 19 pairs, one for each `books` document. For example, one of them will be `("OXF", 1)`. In fact, four of them will be `("MHE", 1)`, because there are four documents in the `books` collection whose `publisher` field is `"MHE"`.

The `reduce1()` function uses the `Array.sum()` method to return the sum of the values of the second argument (`numBooks`) for each value of the first argument (`pupId`). For example, one key-value pair that `reduce1()` will receive as input is `("MHE", [1, 1, 1, 1])`, because the `map1()` function emitted the `("MHE", 1)` pair four times. So, in that case, the array `[1, 1, 1, 1]` is the argument for the parameter `numBooks`, and `Array.sum()` returns 4 for that.

This, of course, is like what we did with the WordCount program (see *Figure 11-2*).

Note that the output from the `mapReduce()` function is a collection, named with a string value assigned to the `out` field; in this case, `"map_reduce_example"` (*Figure 11-6*). So, that data can be accessed, like any other collection, with the `find()` function.

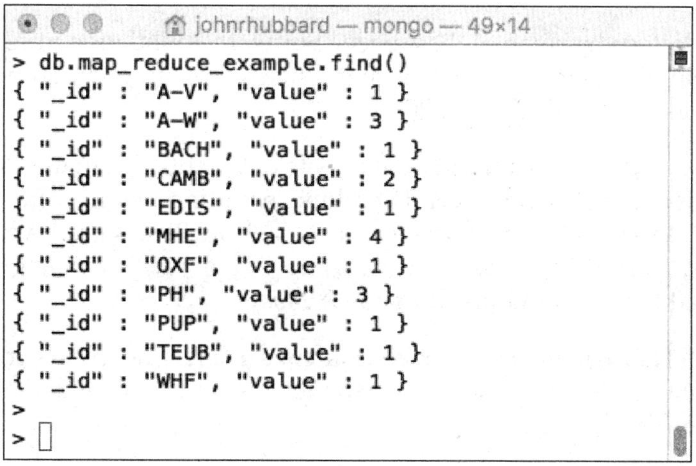

```
> db.map_reduce_example.find()
{ "_id" : "A-V", "value" : 1 }
{ "_id" : "A-W", "value" : 3 }
{ "_id" : "BACH", "value" : 1 }
{ "_id" : "CAMB", "value" : 2 }
{ "_id" : "EDIS", "value" : 1 }
{ "_id" : "MHE", "value" : 4 }
{ "_id" : "OXF", "value" : 1 }
{ "_id" : "PH", "value" : 3 }
{ "_id" : "PUP", "value" : 1 }
{ "_id" : "TEUB", "value" : 1 }
{ "_id" : "WHF", "value" : 1 }
>
>
```

Figure 11-7. Reviewing the output from mongo MapReduce execution

Apache Hadoop

Apache Hadoop is an open-source software system that allows for the distributed storage and processing of very large datasets. It implements the MapReduce framework.

The system includes these modules:

- **Hadoop Common**: The common libraries and utilities that support the other Hadoop modules
- **Hadoop Distributed File System (HDFS™)**: A distributed filesystem that stores data on commodity machines, providing high-throughput access across the cluster
- **Hadoop YARN**: A platform for job scheduling and cluster resource management
- **Hadoop MapReduce**: An implementation of the Google MapReduce framework

Hadoop originated as the Google File System in 2003. Its developer, Doug Cutting, named it after his son's toy elephant. By 2006, it had become **HDFS**, the **Hadoop Distributed File System**.

In April of 2006, using MapReduce, Hadoop set a record of sorting 1.8 TB of data, distributed in 188 nodes, in under 48 hours. Two years later, it set the world record by sorting one terabyte of data in 209 seconds, using a 910-node cluster.

Cutting was elected chairman of the Apache Software Foundation in 2010. It is probably the largest maintainer of free, highly-useful, open source software. Apache Hadoop 2.8 was released in March, 2017.

Hadoop can be deployed in a traditional on-site data center, or in the cloud. Hadoop cloud service is available from several vendors, including Google, Microsoft, Amazon, IBM, and Oracle. For example, The New York Times used Amazon's cloud service to run a Hadoop application, processing 4 TB of raw TIFF image data into 11 million finished PDFs in 24 hours, for about $240.

You can install Hadoop on your own computer as a single-node cluster. This is explained at `https://hadoop.apache.org/docs/stable/hadoop-project-dist/hadoop-common/SingleCluster.html`.

Hadoop MapReduce

After installing Hadoop you can run its version of MapReduce quite easily. As we have seen, this amounts to writing your own versions of the map() and reduce() methods to solve the particular problem. This is done by extending the Mapper and Reducer classes defined in the package org.apache.hadoop.mapreduce.

For example, to implement the WordCount program, you could set your program up like the one shown in *Listing 11-5*.

```java
public class WordCount {

    public static class WordCountMapper extends Mapper {
        public void map(Object key, Text value, Context context) {

        }
    }

    public static class WordCountReducer extends Reducer {
        public void reduce(Text key, Iterable values, Context context) {

        }
    }

    public static void main(String[] args) {

    }
}
```

Listing 11-5. WordCount program in Hadoop

The main class has two nested classes named WordCountMapper and WordCountReducer. These extend the corresponding Hadoop Mapper and Reducer classes, with a few details omitted. The point is that the map() and reduce() methods, that are to be written, are defined in these corresponding classes. This structure is what makes the Hadoop MapReduce framework an actual software framework.

Note that the Text class used in the parameter lists at lines 11 and 17 are defined in the org.apache.hadoop.io package.

This complete example is described at: https://hadoop.apache.org/docs/r2.8.0/hadoop-mapreduce-client/hadoop-mapreduce-client-core/MapReduceTutorial.html

Summary

This chapter presents some of the ideas and algorithms involved in the analysis of very large datasets. The two main algorithms are Google's PageRank algorithm and the MapReduce framework.

To illustrate how MapReduce works, we have implemented the WordCount example, which counts the frequencies of the words in a collection of text files. The more realistic implementation would be with MongoDB, which is presented in *Chapter 10, NoSQL Databases*, or in Apache Hadoop, which is briefly described in this chapter.

Java Tools

This appendix gives a brief description and installation instructions about the various software tools that are used in the book. Each of these tools is free and can be installed fairly easily on your own computer, whether you are running macOS X, Microsoft Windows, or some variety of UNIX. Since everything in this book was written on a Mac, we will focus on that platform. Installations and maintenance on other platforms is similar.

This information is current as of August, 2017. The online links are likely to be updated periodically, but the basic steps for installation and use are not expected to change much.

The command line

On a Mac, the command line is accessed through the Terminal app (in Windows, it's called **Command Prompt**.) You will find the Terminal app in the `Applications/Utilities/` folder. When you launch it, a Terminal window will appear, like the one in *Figure A-1*:

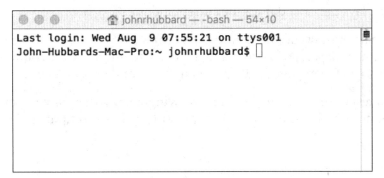

Figure A-1. Terminal window

The prompt shows the name of the computer, a colon (:), the current directory (the folder), a space, the current user, a dollar sign ($), a space, and then the prompt symbol (▯).

There are several hundred commands that you can run from the Terminal window. To see a list of them all, hold the *Esc* key down for a second and then press *Y*, to answer yes to the question. Each time the listing pauses, press the space bar to see the next screen of commands. Press *Q* to terminate the listing.

Press *Ctrl+C* to abort the execution of any Terminal command. Use the up and down arrow keys to scroll through the saved list of previously executed commands (to avoid re-typing them).

Try the `cal` command (*Figure A-2*):

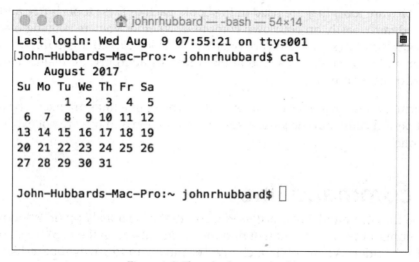

Figure A-2. The calendar command

It simply displays the calendar for the current month.

Most Terminal window commands are Unix commands. To view the Unix manual description (the *man page*) of a command, type `man` followed by the command. For example, to see the man page for the calendar command, enter `man cal`.

Whenever a response is too large to fit in the window, use the space bar or the *Q* key to either continue to the next screen of output or to quit that output.

Appendix

The prompt string is saved in a system variable name `PS1`. You can change it to just about anything you want by a simple assignment statement. Here we changed it, first to `"Now what?"` (just for fun), and then to `"\w $"`. The code `\w` means to specify the current working directory, representing the user's home directory with the tilde character (~). The current working directory in *Figure A-3* is the user's home directory.

```
John-Hubbards-Mac-Pro:~ johnrhubbard$ PS1="Now what? "
[Now what? cal
      August 2017
Su Mo Tu We Th Fr Sa
       1  2  3  4  5
 6  7  8  9 10 11 12
13 14 15 16 17 18 19
20 21 22 23 24 25 26
27 28 29 30 31

[Now what? say Just who do you think you are?
[Now what? PS1="\w $ "
~ $
```

Figure A-3. Changing the Command Prompt string

The part of the computer's operating system that responds to user commands this way is called the **shell** and a series of interactive commands and responses like these is called a **shell session**.

The shell session shown in *Figure A-4* illustrates the `cd` and `ls` commands. The `cd` command changes the current directory to the one specified in the command. So, `cd hub` changed the current directory from `~` (my home directory) to `~/hub`, which is a subdirectory of my home directory.

```
[Now what? PS1="\w $ "
[~ $ cd hub
[~/hub $ ls
app      data     im       net      pro      ur
books    gen      misc     per      tmp
[~/hub $ cd ur
[~/hub/ur $ ls
admin         misc              research
courses       orchestra
[~/hub/ur $ cd courses
[~/hub/ur/courses $ ls
cs150    cs222    cs325    cs395
cs221    cs315    cs340
~/hub/ur/courses $
```

Figure A-4. Listing contents and changing directories

The `ls` command lists the contents of the current directory. So, when I executed the `ls` command the first time here, it listed the 11 subdirectories that I have in my `~/hub` directory. These included `app`, `books`, `data`, and so on.

To go back up your directory tree, use the `cd ..` command, or `cd ../..` to go back up two levels at once. The double dot means the parent directory.

Java

Java has been, for at least a decade now, the most popular programming language. Nearly all the software described in this appendix is itself written in Java.

Java was developed in the early 1990s by a team of software researchers, led by James Gosling. It was released by Sun Microsystems in 1995 and was maintained by that company until it was acquired by Oracle in 2010.

Java comes pre-installed on the Mac. To update it, open **System Preferences** and click on the **Java** icon at the bottom of the panel. This brings up the **Java Control Panel**. The **About...** button under the **General** tab will display the current version of Java that is installed. As of August, 2017, that should be Version 8. The **Update** tab allows for updates.

Java can be run from the command line or from within an **Integrated Development Environment** (**IDE**). To do the former, you will need a text editor to write your code. On the Mac, you can use the TextEdit app. The four-line Hello World program is shown in a TextEdit window in *Figure A-5*.

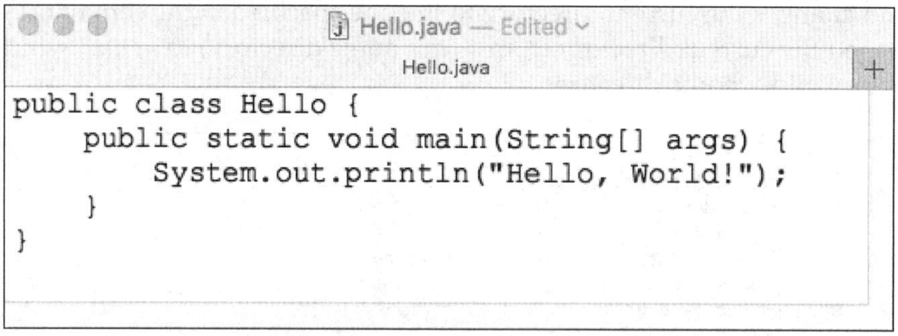

```
public class Hello {
    public static void main(String[] args) {
        System.out.println("Hello, World!");
    }
}
```

Figure A-5. The Hello World program in Java

Appendix

Note that, for this to compile, the quotation marks must be flat (like this " "), not smart (like this " "). To get that right, open the **Preferences** panel in TextEdit and deselect **Smart Quotes**. Also note that the TextEdit's default file types, RTF and TXT, must be overridden—Java source code must be saved in a file whose name is the same as the `public` class and whose type is `.java`.

We saved the source code in our `hub/demos/` folder. The shell session in *Figure A-6* appears as follows:

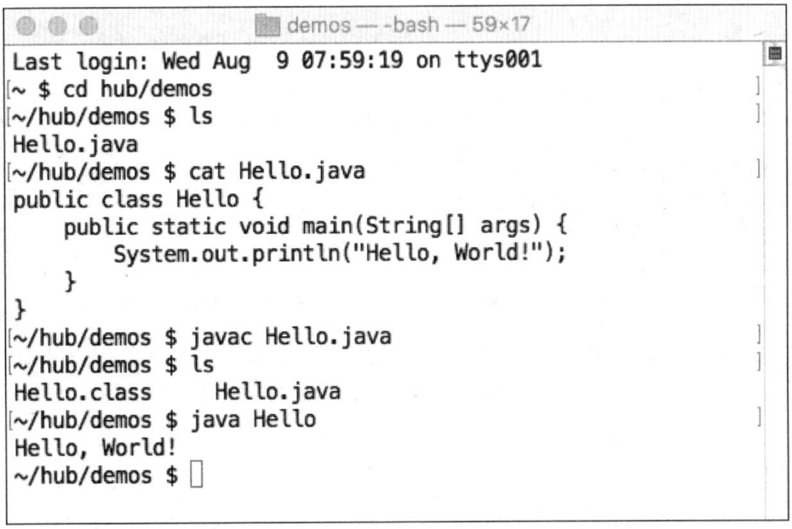

```
Last login: Wed Aug  9 07:59:19 on ttys001
[~ $ cd hub/demos
[~/hub/demos $ ls
Hello.java
[~/hub/demos $ cat Hello.java
public class Hello {
    public static void main(String[] args) {
        System.out.println("Hello, World!");
    }
}
[~/hub/demos $ javac Hello.java
[~/hub/demos $ ls
Hello.class    Hello.java
[~/hub/demos $ java Hello
Hello, World!
~/hub/demos $ 
```

Figure A-6. Running the Hello World program

From this, we can learn:

- How to display the source code with the `cat` command
- How to compile the source code with the `javac` command
- How to run the program with the `java` command

The `javac` command creates the class file, `Hello.class`, which we then execute with the `java` command, thus generating the expected output.

Java Tools

The `Echo` program in *Figure A-7* illustrates the use of command line arguments.

```
~/hub/demos $ ls
Echo.java         Hello.class      Hello.java
~/hub/demos $ cat Echo.java
public class Echo {
    public static void main(String[] args) {
        for (int i = 0; i < args.length; i++) {
            System.out.printf("args[%d] = %s%n", i, args[i]);
        }
    }
}
~/hub/demos $ javac Echo.java
~/hub/demos $ java Echo alpha beta gamma delta
args[0] = alpha
args[1] = beta
args[2] = gamma
args[3] = delta
~/hub/demos $
```

Figure A-7. Reading command line arguments

The program prints each element of its `args[]` array. In the run shown here, there are four of them, entered on the command line immediately after the name of the program.

NetBeans

NetBeans is the IDE that we have used throughout this book. As mentioned in *Chapter 1, Introduction to Data Analysis*, it is comparable to the other popular IDEs, such as Eclipse, JDeveloper, and JCreator, and it functions mostly the same way.

Download NetBeans from `https://netbeans.org/downloads/`. Select your language, your OS platform, and one of the **Download** buttons. Unless your machine is short of storage space, you might as well select the **All** (right most **Download** button) version. It includes the Java Enterprise Edition (Java EE) along with support for HTML5 and C++.

The installation is straightforward: just follow the directions. Note that NetBeans includes Java, so you do not have to install Java separately.

Appendix

The main window for NetBeans is shown in *Figure A-8*:

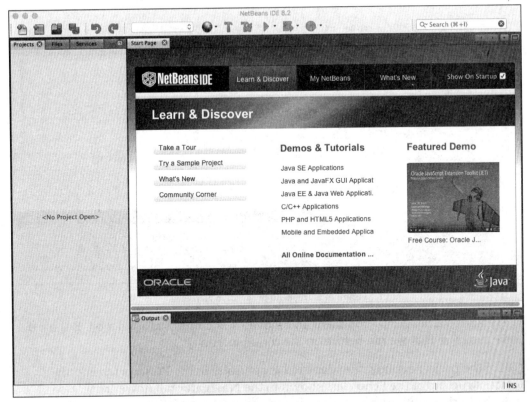

Figure A-8. The NetBeans IDE

It shows five main panels: **Projects**, **Files**, **Services**, **Output**, and the editing window. The **Projects**, **Files**, and **Services** panels are combined in the sub-window on the left. This configuration can easily be rearranged.

NetBeans wants you to compartmentalize all your Java work within the context of NetBeans projects. Although it may seem like an unnecessary extra step, it makes it much easier in the long run to keep track of all your work.

To create a project, just click on the New Project button ( ) on the toolbar (or select **New Project...** from the **File** menu). Then, in the **New Project** dialogue, select **Java | Java Application**, choose your **Project Name**, **Project Location**, and **Main Class** (for example, `AppendixA`, `nbp/`, and `dawj.appA.Echo`), and click **Finish**. Note that the **Main Class** name includes a package specification (`dawj.appA`).

After pasting in our `Echo` program, the setup looks like *Figure A-9*:

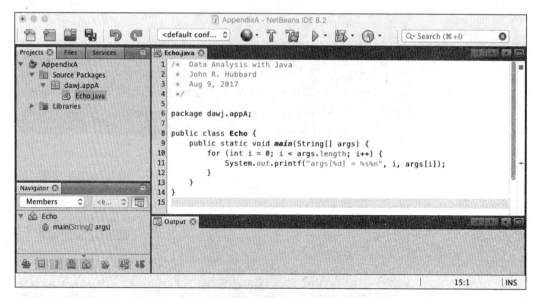

Figure A-9. The Echo program in NetBeans

We've added a header ID comment and moved some things around a bit. But it's the same program that we ran before on the command line.

Notice the project organization depicted graphically in the **Projects** tab and the logical structure of the **Echo** class shown in the **Navigator** tab (lower left). Also note the varied iconography: the golden coffee cup for a NetBeans project, the tan bundle cube for the package folder, the icon with the tiny green triangle indicating an executable (main) class, and the orange disk with a vertical bar indicating a static method. These clues can be quite helpful when the project grows to include many classes and methods.

To run the program, click on the green triangle button (▷) on the Toolbar (or press the *F6* key, or select **Run Project** from the **Run** menu).

Note that selecting the button will run the project. If your project has more than one class with a `main()` method, it will have designated one of them as its main class and will run that one. To run a program that is not so designated, select **Run File** from the **Run** menu, or press *Shift+F6*. To re-designate the main class for a project, right-click on the project's icon, and select **Properties** | **Run**.

Notice that this `Echo` program gets all its input from the command line. Without that input, it will produce no output. To provide command line arguments, right-click on the project's icon in the **Projects** panel, select **Properties** | **Run**, and then insert them in the **Arguments**: field, just as you would on the command line (without any punctuation). See *Figure A-10*.

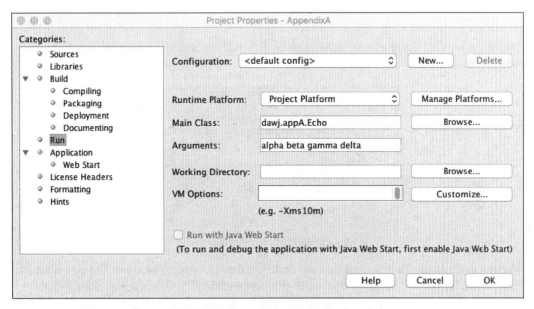

Figure A-10. The NetBeans project properties Run dialogue

Then, when you run it, you will get the expected output.

The NetBeans editor is very powerful. It uses different colors to distinguish reserved words (blue), class fields (green), and strings (orange) from other code. Syntax errors are caught immediately and marked with red tags. Recommendations are provided for correcting most errors. If you use a class, the IDE will give you a list of import statements from which to choose. If you right-click on a class or class member (field or method) name, you can select **Show Javadoc** from the pop-up menu and it will appear in your browser. You can select any part of your code and have it properly formatted with *Ctrl+Shift+F* (or right-click and choose **Format**).

The built-in debugger is also very powerful and very easy to use.

NetBeans comes with its own relational database system, which is illustrated in *Chapter 5, Relational Databases*.

NetBeans is excellent for developing Java software. It is also supports C, C++, Fortran, HTML5, PHP, JSP, SQL, JavaScript, JSON, and XML.

MySQL

To download and install MySQL Community Server, go to `https://dev.mysql.com/` and select a recent version of MySQL Community Server. Select your platform (for example, macOS X) and click on the **DMG Archive** or the **TAR Archive**. Then run the installer. When the installation has completed, you may have to restart your system.

The installer will issue a temporary password to the access the database server (see *Figure A-11*):

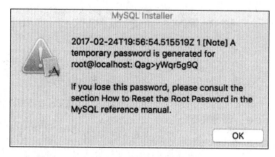

Figure A-11. The MySQL Installer

Make a note of the 12-character password (the one in your panel, not the one shown here). You'll need it shortly. Also, note the connection ID: `root@localhost`. That means that your username is `root` and your computer's name is `localhost`.

After installing the server, it has to be started. To do this on the Mac, open **System Preferences**. At the bottom of the panel, you will see the MySQL icon. Click on it to open the MySQL server panel (see *Figure A-12*).

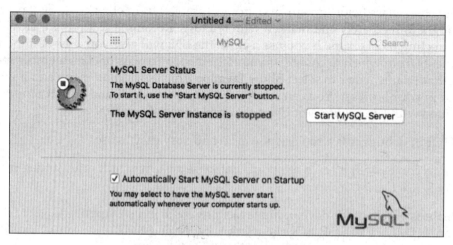

Figure A-12. Starting the MySQL server

Also, check that the **Automatically Start MySQL Server on Startup** option is enabled.

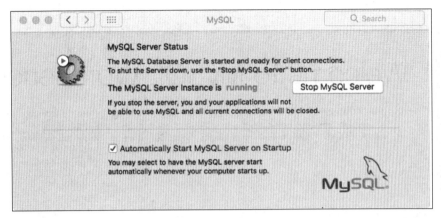

Figure A 13. Stopping the MySQL server

MySQL Workbench

We will be creating and accessing our MySQL databases by two different methods: by running our own independent Java programs and through a user interface tool named **MySQL Workbench**.

To download and install MySQL Workbench, go back to https://dev.mysql.com/downloads/ and select **MySQL Workbench**. Select your platform and click on the **Download** button. Then, run the installer.

In the window that comes up, drag the **MsSQLWorkbench.app** icon to the right, into the `Applications` folder:

Figure A-14. Installing MySQL Workbench

After a few seconds, the app's icon will appear in the `Applications` folder. Launch it, and affirm that you want to open it:

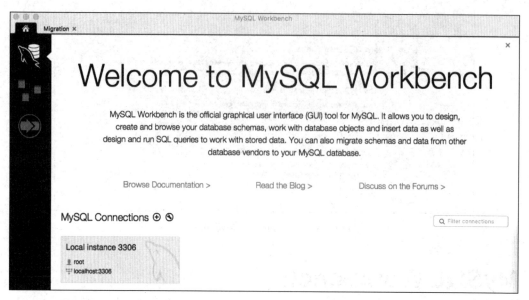

Figure A-15. Running the MySQL Workbench installer

Note that one connection is defined: `localhost:3306` with the user root. The number `3306` is the port through which the connection is made.

Now that we have MySQL Workbench installed, we can test it. The main window for the app is shown in *Figure A-16*.

Appendix

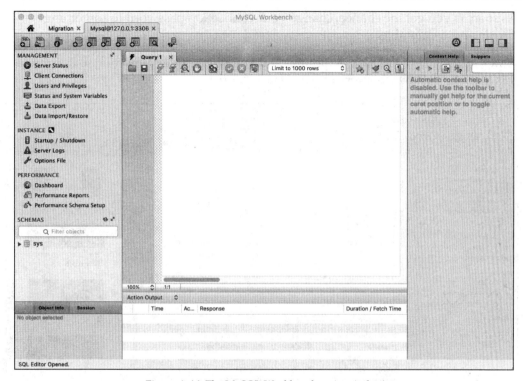

Figure A-16. The MySQL Workbench main window

Notice the tab labeled `Mysql@127.0.0.1:3306`. That is the name of the connection that we have defined (here, `127.0.0.1` is the IP address of `localhost`). To actually connect to the database through that service, we must open the connection.

In the MySQL Workbench menu, select **Database | Connect to Database…**. Enter the password that you saved and click **OK** (see Figure A-17).

Figure A-17. Connecting to the MySQL Server

[367]

Java Tools

As soon as you get the chance, change your connection password to something simple and memorable.

Now your MySQL database is running and you have a live connection to it through this MySQLWorkbench interface (see *Figure A-18*).

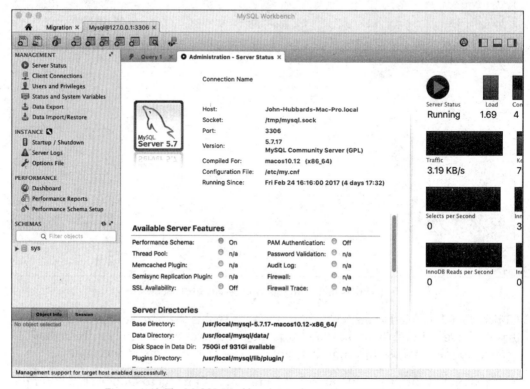

Figure A-18. The MySQL Workbench interface to the MySQL database

To create a database, click on this icon in the toolbar: This is the **New Schema** button. The image, which looks like a 55 gallon barrel of oil, is supposed to look like a big disk, representing a large amount of computer storage. It's a traditional representation of a database.

A new tabbed panel, labeled `new_schema`, opens. Technically, we are creating a schema on the MySQL Server.

Enter `schema1` for **Schema Name**:

Figure A-19. Creating a database schema in MySQL

Then, click on the **Apply** button. Do the same on the **Apply SQL Script** to **Database** panel that pops up and then close that panel.

You now have two schemas listed under **SCHEMAS** in the sidebar: **schema1** and **sys**. Double-click on **schema1** to make it the current schema. That expands the listing to show its objects: **Tables**, **Views**, **Stores Procedures**, and **Functions** (see *Figure A-20*).

Figure A-20. schema1 objects

Java Tools

We can think of this as our current database. To add data to it, we must first create a table in it.

Click on the tab labeled **Query 1** and then type the code shown in *Figure A-21* into the editor.

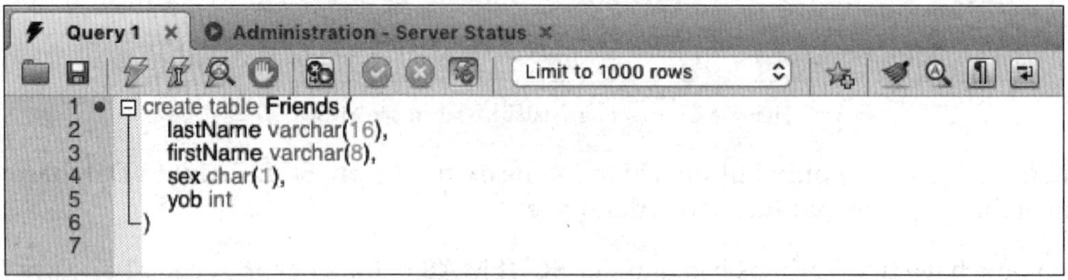

Figure A-21. SQL code

This is SQL code (SQL is the query language for RDBs). When this query is executed, it will create an empty database table named `Friends` that has four fields: `lastName`, `firstName`, `sex`, and `yob`. Notice that in SQL, the datatype is specified after the variable that it declares. The `varchar` type holds a variable-length character string of a length from 0 up to the specified limit (16 for `lastName` and 8 for `firstName`). The `char` type means a character string of exactly the specified length (1 for `sex`) and `int` stands for integer (whole number).

Note that SQL syntax requires a comma to separate each item in the declaration list, which is delimited by parentheses.

Also note that this editor (like the NetBeans editor) color-codes the syntax: blue for keywords (`create`, `table`, `varchar`, `char`, `int`) and orange for constants (16, 8, 1).

SQL is a very old computer language. When it was introduced in 1974, many keyboards still had only upper-case letters. So, it became customary to use all capital letters in SQL code. That tradition has carried over and many SQL programmers still prefer to type keywords in all caps, like this:

```
CREATE TABLE Friends (
    lastName VARCHAR(16),
    firstName VARCHAR(8),
    sex CHAR(1),
    yob INT
)
```

But it's really just a matter of style preference. The SQL interpreter accepts any combination of uppercase and lowercase letters for keywords.

To execute this query, click on the yellow lightning bolt on the tabbed panel's toolbar:

Expand the `schema1` tree in the sidebar to confirm that your `Friends` table has been created as intended (see *Figure A-22*):

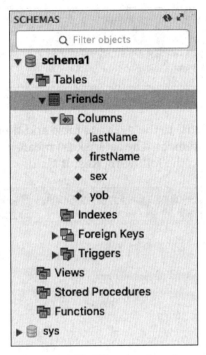

Figure A-22. Objects in schema1

Now we're ready to save some data about our friends.

Click on the new query button (or select **File | New Query Tab**). Then execute the query shown in *Figure A-23*. This adds one row (data point) to our `Friends` table. Notice that character strings are delimited by the apostrophe character, not quotation marks.

Java Tools

Next, execute the queries shown in *Figure A-23*, *Figure A-24*, and *Figure A-25*:

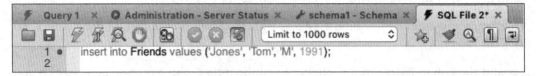

Figure A-23. An insertion

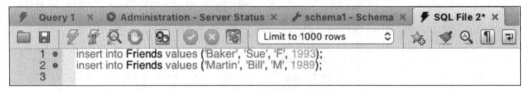

Figure A-24. Two more insertions

These add three rows (records) to the `Friends` table and then queries the table. The query is called a `select` statement. The asterisk (*) means to list all the rows of the table. The output is shown in the following **Result Grid**:

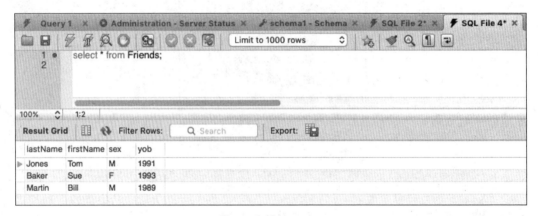

Figure A-25. A query

Appendix

Accessing the MySQL database from NetBeans

To access the MySQL database from NetBeans, click on the **Services** tab in the NetBeans left sidebar and expand the **Databases** section:

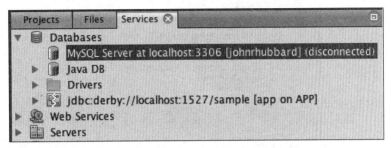

Figure A-26. Locating the MySQL database in NetBeans

Then, right-click the MySQL Server icon and select **Connect**:

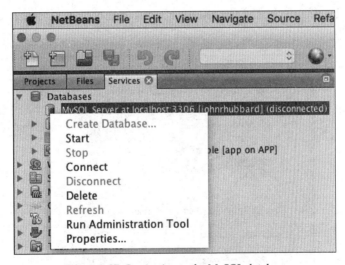

Figure A-27. Connecting to the MySQL database

When asked about editing your MySQL Server connection properties, click **Yes**.

Use your MySQL username and password to connect to the database. Then, expand the MySQL Server node to see the accessible schema, as shown in *Figure A-28*:

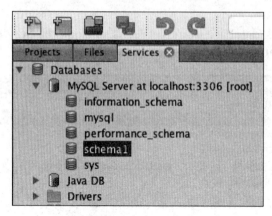

Figure A-28. Connecting to schema1

Then, right-click on **schema1** and then on **Connect** in the pop-up panel.

Next, expand the node whose icon shows a solid (not broken) connection, as shown in *Figure A-29*:

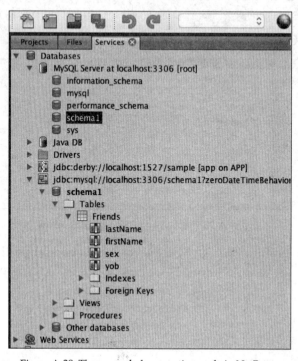

Figure A-29. The expanded connection node in NetBeans

Appendix

This shows the details of the `Friends` table that you created and loaded in MySQL Workbench.

Right-click again on the **jdbc:mysql** node and select **Execute Command…**, as shown in *Figure A-30*.

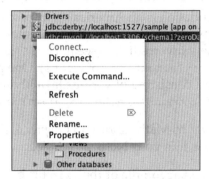

Figure A-30. Executing a SQL command

This opens a new tabbed pane, labeled **SQL 1 [jdbc:myscl://localhost:33…]**.

Now, type the same SQL `select` query that you used before in MySQL Workbench. (see *Figure A-25*). The output is displayed at the bottom of the window, as shown in *Figure A-31*:

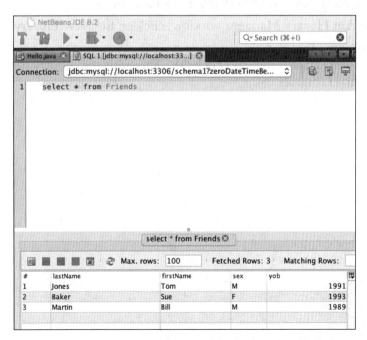

Figure A-31. Executing a query on a MySQL database from NetBeans

Java Tools

This shows that SQL access to a MySQL database in NetBeans is very similar to that in MySQL Workbench.

The Apache Commons Math Library

The Apache Software Foundation is an American non-profit corporation that supports open source software projects written by volunteers. It includes a vast number of projects in widely diverse fields (see `https://en.wikipedia.org/wiki/List_of_Apache_Software_Foundation_projects`). One part of this collection is called Apache Commons, consisting of various Java libraries. These can be downloaded from `http://commons.apache.org/downloads/`.

We used the Apache Commons Math Library in *Chapter 6, Regression Analysis*. Here are the steps to follow to use it in your NetBeans projects:

1. Download either the `tar.gz` file or the `.zip` file of the most recent version (3.6.1, as of August, 2017) of the commons-math archive file from: `http://commons.apache.org/proper/commons-math/download_math.cgi`.
2. Expand the archive (double-click or right-click on it), and then copy the resulting folder to a location where other Java libraries are kept on your machine (for example, `Library/Java/Extensions`).
3. In NetBeans, select **Libraries** from the **Tools** menu.
4. In the **Ant Library Manager** window, click on the **New Library...** button.
5. In the **New Library** dialogue, enter `Apache Commons Math` for the **Library Name** (see *Figure A-32*), and click **OK**. This adds that library name to your NetBeans **Libraries** list.

Appendix

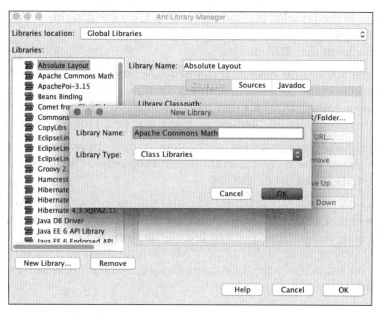

Figure A-32. Creating a new library in NetBeans

6. Next, with the **Classpath** tab selected, click the **Add JAR/Folder...** button and then navigate to the folder where you copied the folder in *step 2*. Select the `commons-math3-3.6.1.jar` file inside that folder (see *Figure A-33*) and click on the **Add JAR/Folder button**.

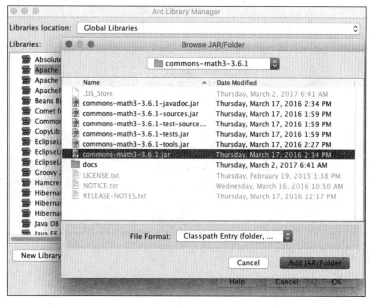

Figure A-33. Locating the JAR file for a NetBeans library

7. Switch to the **Javadoc** tab and repeat *step 6*, except this time, click the **Add ZIP/Folder...** button and then select the `commons-math3-3.6.1-javadoc.jar` file from that same folder.

8. Then click **OK**. You have now defined a NetBeans library named Apache Commons Math, containing both the compiled JAR and the Javadoc JAR for the **commons-math3-3.6.1** that you downloaded. Henceforth, you can easily add this library to any NetBeans project that you'd like to have use it.

9. To designate that library for use by your current (or any other) project, right-click on the project icon and select **Properties**. Then select **Libraries** in the **Categories** list on the left.

10. Click on the **Add Library...** button, and then select your **Apache Commons Math** library from the list. Click on the **Add Library** button and then **OK** (see *Figure A-34*). Now, you can use all the items in that library in any source code that you write in that project. Moreover, the Javadocs should work the same for those packages, interfaces, classes, and members as they do for standard Java code.

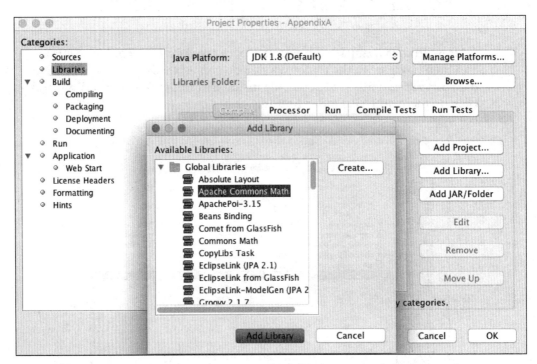

Figure A-34. Adding the Apache Commons Math Library to a specific NetBeans project

Appendix

To test your installation, do this:

1. In your main class, add this code:

   ```
   SummaryStatistics stats;
   ```

2. Since the `SummaryStatistics` class is not part of the standard Java API, NetBeans will mark that line as erroneous, like this:

   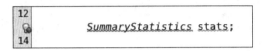

3. Click on the tiny red ball in the current line's margin. A drop-down list appears (see *Figure. A-35*).

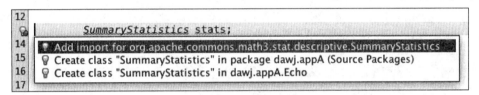

Figure A-35. Adding the correct import statement in NetBeans

4. With the Add import item selected (the first line in in the list), press *Enter*. That will cause the import statement to be inserted in your source code (line 7, *Figure. A-36*).

```
 5  package dawj.appA;
 6
 7  import org.apache.commons.math3.stat.descriptive.SummaryStatistics;
 8
 9  public class Echo {
10      public static void main(String[] args) {
11          for (int i = 0; i < args.length; i++) {
12              System.out.printf("args[%d] = %s%n", i, args[i]);
13          }
14
15          SummaryStatistics stats;
16      }
17  }
```

Figure A-36. Getting the import statement inserted automatically in NetBeans

5. Next, click on the `SummaryStatistics` class name where the `stats` variable is being declared, and then select **Show Javadoc** from the drop-down menu that appears. That should bring up the Javadoc page for that class in your default web browser.

Java Tools

6. Once you have any Javadoc page displayed from a library (such as **org.apache.commons.math3**), you can easily investigate all the other subpackages, interfaces, and classes in that library. Just click on the **Frames** link at the top of the page and then use the sliding lists in the two frames on the left (*Figure A-37*).

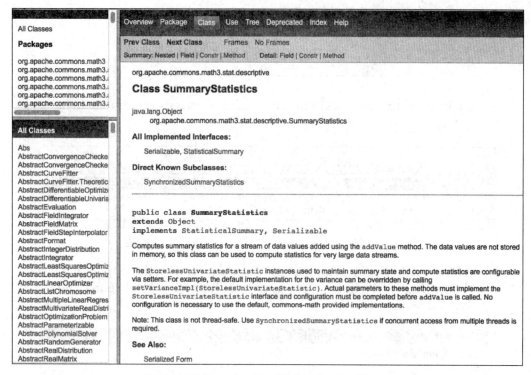

Figure A-37. Surveying the other Javadocs in a library

The javax JSON Library

The `javax.json` library, used in *Chapter 2, Data Preprocessing*, is actually included in the existing NetBeans libraries. It is part of Java EE (the Enterprise Edition). If you installed one of the versions of NetBeans that includes Java EE, then you don't have to download and install the JSON library separately.

You will need to add that library to your NetBeans project though. Just repeat steps 9-10 from *The Apache Commons Library* section for the Java EE 7 API Library (see *Figure A-34*).

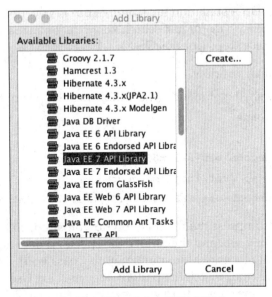

Figure A-38. Adding the Java EE 7 API Library

The Weka libraries

The Weka platform of data analysis implementations is maintained by computer scientists at the University of Waikato in New Zealand. It includes some Java libraries that we have used in this book. For example, the `TestDataSource` program, shown in *Listing 7-5*, uses several classes from the `weka.core` package.

Download Weka from http://www.cs.waikato.ac.nz/ml/weka/downloading.html. The bundle includes the Weka app itself and a folder (`weka-3-9-1`, as of August, 2017) of Java libraries. Inside the folder you will find the `weka.jar` file and a folder named `doc`.

To add the Weka library to NetBeans, follow the same *steps 1-8* described previously for adding the Apache Commons Math library, except with `weka.jar` instead of `commons-math3-3.6.1.jar`. You can name the library `Weka 3.9.1` (or anything you like). For **Classpath**, add the `weka.jar` file. For **Javadoc**, add the `doc` folder. Then you can add that Weka 3.9.1 library to any NetBeans project the same way that the Apache Commons Math library was added in *steps 9-10*.

Java Tools

MongoDB

The MongoDB NoSQL database system is described in *Chapter 10, NoSQL Databases*.

Download the MongoDB installation file from `https://www.mongodb.com/download-center#community` (or just search `MongoDB download` to find that web page). Then, follow these steps (for macOS X):

1. Open a Terminal window.
2. Execute these shell commands (the $ indicates the prompt):

   ```
   $ cd ~/Downloads
   $ ls mongo*
   $ tar xzf mongodb-osx-ssl-x86_64-3.4.7.tar
   $ sudo mkdir -p /usr/local/mongodb
   $ sudo mv mongodb-osx-x86_64-3.4.7 /usr/local/mongodb
   ```

   These will decompress the file that you downloaded, create the folder `/usr/local/mongodb`, and then move the specified decompressed file to that folder. You may be asked for your computer password. The `ls` command will list the files that match the pattern `mongo*.tar`. Use that to get the right filename. (The one shown here was correct in August, 2017.)

3. Execute these commands next:

   ```
   $ sudo mkdir -p /data/db
   $ sudo chown johnrhubbard /data/db
   ```

   Use your own username with the `chown` command. This creates the directory `/data/db` and gives you ownership of it.

4. Execute these commands next:

   ```
   $ cd ~
   $ touch .bash_profile
   $ vim .bash_profile
   ```

   This starts the vim editor, editing your `.bash_profile` file. All the commands that you save in that file will be executed automatically whenever you start a shell session.

5. In the `vim` editor, do this:
    1. Type `i` to shift into the insert mode.
    2. Add these three lines to the file:
        ```
        export MONGO_PATH=/usr/local/mongodb/mongodb-osx-x86_64-3.4.7
        export PATH=$PATH:$MONGO_PATH/bin
        PS1="\w $ "
        ```
    3. Make sure you get the `mongodb` file name right.
    4. Press the *Esc* key to exit the insert mode.
    5. Then type ZZ to save the file and exit vim.

    The first line defines the `MONGO_PATH` variable. Note that the first line is shown here on two lines. The second line adds that path to your `PATH` variable. The third line simply sets your `PS1` prompt (this was discussed at the beginning of this chapter.) It's optional and you can set it to anything you want.

6. Execute this command next:
    ```
    $ source .bash_profile
    ```
    The `source` command executes the specified file. If you reset your `PS1` prompt, you will see the result.

7. Finally, to check that Mongo is installed, execute this command:
    ```
    $ mongo -version
    ```

The response should confirm the installation by reporting on the version installed.

Now that the system has installed, you are ready to start it. Do this in a new shell window (just type *Ctrl+N* to get a new window.) Execute this command to start the MongoDB database system:

```
$ mongod
```

This will produce many rows of output.

It may ask whether you want the application mongod to accept incoming network connections. Unless you have security concerns, click the **Allow** button.

Set that window aside. You can hide it, but don't close it.

Now to start a MongoDB shell, where you can execute MongoDB commands, open another Terminal window, and execute this command:

```
$ mongo
```

After about 10 lines of output, you should finally see the Mongo shell prompt:

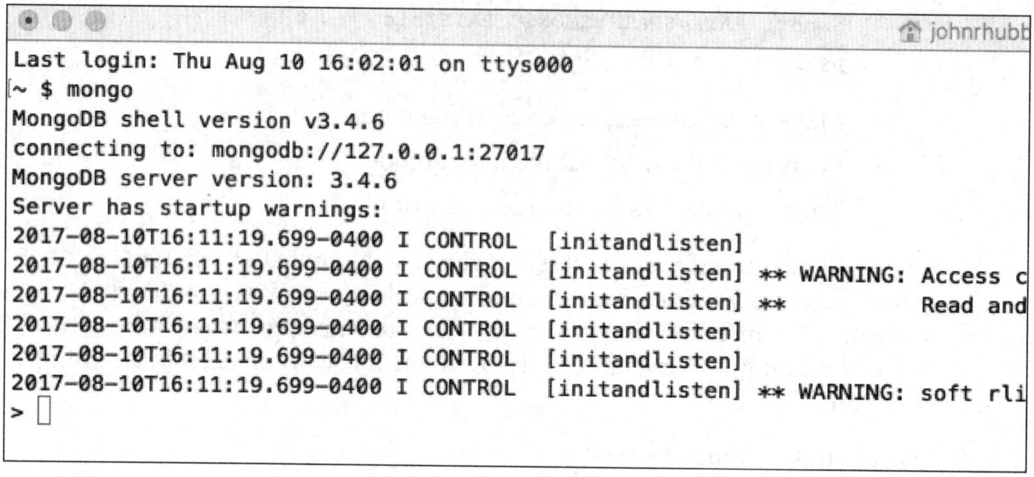

```
Last login: Thu Aug 10 16:02:01 on ttys000
[~ $ mongo
MongoDB shell version v3.4.6
connecting to: mongodb://127.0.0.1:27017
MongoDB server version: 3.4.6
Server has startup warnings:
2017-08-10T16:11:19.699-0400 I CONTROL  [initandlisten]
2017-08-10T16:11:19.699-0400 I CONTROL  [initandlisten] ** WARNING: Access c
2017-08-10T16:11:19.699-0400 I CONTROL  [initandlisten] **          Read and
2017-08-10T16:11:19.699-0400 I CONTROL  [initandlisten]
2017-08-10T16:11:19.699-0400 I CONTROL  [initandlisten]
2017-08-10T16:11:19.699-0400 I CONTROL  [initandlisten] ** WARNING: soft rli
>
```

Figure A-39. The Mongo shell prompt

At this point, you can continue with the content in *Chapter 10, NoSQL Databases*.

Index

Symbol

95% confidence coefficient 103, 104

A

actuarial science 3
affinity propagation clustering 256-264
aggregate function 139
alternative hypothesis 105
Amazon Standard Identification Number (AISM) 306
American College Board Scholastic Aptitude Test in mathematics (AP math test) 96
Anscombe's quartet 163, 164
Apache Commons
  download link 376
Apache Commons implementation 174, 175
Apache Commons Math Library
  about 376
  using 376-380
Apache Hadoop 352
Apache Software Foundation
  about 376
  references, for projects 376
ARFF filetype 198
associative array 303
Attribute-Relation File Format (ARFF)
  about 198, 200
  reference 198
attributes 14
attribute-value pairs 16

B

backward elimination 224

bar chart
  about 49
  generating 49
batch mode 123
batch processing 130-133
Bayesian classifiers
  about 203-205
  Java implementation, with Weka 206-209
  support vector machine algorithms 209
Bayes' theorem 91-93
big data 331
Binary Search algorithm 180
binomial coefficient 83, 84
binomial distribution 84-87
bins 51
BSON object 319
B-tree 140
bulk write 312

C

Canberra metric 231
Cartesian plane 228
Cartesian product 109
Cassandra 330
central limit theorem 102, 103
Chebyshev metric 230
chessboard metric 231
clustering algorithm 227
column data model 330
column store 311
comma separated values (CSV) 19
Commons Math API
  reference 76
complexity analysis 243
compound indexes 329

conditional probability 89, 90
confidence intervals 103, 104
content-based recommendation 269
contingency tables 91
continuous distribution 97
correlation coefficient 93
cosine similarity 271, 272
covariance 93-96
crosstab tables 91
cumulative distribution
    function (CDF) 68, 81-83
curse of dimensionality 233-235
curve fitting 176, 177

D

data
    about 8
    inserting, into database 121-123
data analysis
    about 1
    origins 1, 2
database
    creating 114-117
database driver 125
database query 123
database schema
    creating, in MySQL 371, 372
database views 134-137
data cleaning 32
Data Definition Language (DDL) 117
data filtering 33-36
Data Manipulation Language (DML) 117
data normalization 32
data points 14
data ranking 61, 63
data scaling 32
data scrubbing 32
datasets 14
data striping 332
data types 13
decile 74
decision tree 180, 181
density function 65
descriptive statistic 73, 74
dictionary 303
discrete distribution 97

distances
    measuring 227-232
divide and conquer 338
document data model 329
document store 311
domain 109
dynamic schemas 320

E

Electronic Numerical Integrator and
    Computer (ENIAC) 6
entropy
    defining 181-185
event 79
Excel
    linear regression, performing in 144-148
exemplars 256
explained variation 154
exponential distribution 67, 68
Extensible Markup Language (XML) 24, 25
extrapolation 143

F

false negative 92
false positive 92
fields 14, 110
file formats 18-20
foreign key 111, 112
frequency distribution 63-65
fuzzy classification algorithms 225

G

Generalized Markup Language (GML) 24
GeoJSON object types 327
geospatial databases
    reference 327
Googleplex 331
graph data model 330
graphs 46

H

Hadoop Common 352
Hadoop Distributed File
    System (HDFS) 352

Hadoop MapReduce
  about 352, 353
  reference, for example 353
  WordCount program 353
Hadoop YARN 352
Hamming distance 270
Hamming similarity 270
hash 41
hash codes 42
hash function
  properties 41
hashing 41
hash table 16-18, 295
HBase 330
Herman Hollerith 5, 6
hierarchical clustering
  about 235-243
  affinity propagation clustering 256-264
  k-means clustering 248-253
  k-medoids clustering 253-256
  Weka implementation 245-247
histogram 50
horizontal scaling 332
hypothesis testing 105-107

I

ID3 algorithm
  about 185-195
  Java implementation 195-197
  Java implementation, with Weka 202, 203
indexing
  in MongoDB 328, 329
information 8
instance 110
International Business Machines
    Corporation (IBM) 6
International Standard Book Numbers 305
item-based recommendation 269
item-to-item collaborative filtering
    recommender 284-289
Iterative Dichotomizer 3 185

J

Java
  about 8
  features 8

Java Database Connectivity (JDBC) 125
Java DB 115
Java development
  with MongoDB 318-326
Java implementation
  example 53-56
  of ID3 algorithm 195-197
  of linear regression 155-163
Java Integrated Development
    Environments 9, 10
JavaScript Object Notation (JSON) 25, 26
javax.json library 380
JDBC PreparedStatement
  using 128
joint probability function 87
JSON data 24
JSON event types
  identifying 27
JSON files
  parsing 28-30
JSON (JavaScript Object Notation) 310

K

key field 15
key-value data model 330
key-value pairs (KVP) 16
key values 16
K-Means++ algorithm 248
K-means clustering 248-253
k-medoids clustering 253-256
K-Nearest Neighbor (KNN) 220-224, 248
knowledge 8
kurtosis 76

L

large sparse matrices 294-297
least-squares parabola 166
level of significance 105
lexicographic order 297
Library database 314-317
linear regression
  about 143, 144
  Anscombe's quartet 163, 164
  in Excel 144-148
  Java implementation 155-163

line graph
    about 48
    generating 48
logarithmic time 140
logistic function 216
logistic regression
    about 214, 216
    example 217-220
    K-Nearest Neighbors 220-224
logit function 216
LU decomposition 170

M

Manhattan metric 230
map 295
Map data structure 303, 304
MapReduce
    in MongoDB 350, 351
    matrix multiplication,
        implementing 346-350
MapReduce applications
    examples 339
MapReduce framework 338
marginal probabilities 89
Markov chain 334
matrix multiplication
    with MapReduce 346-350
maximum 73
mean average 73
median 74
merging 38-41
message-passing 256
meta-algorithm 180
metadata 31
method of least squares 166
metric 227
metric space 227
Microsoft Excel
    moving average, computing 59, 60
Microsoft Excel data 20-23
Minard's map
    of Napoleon's Russian campaign 45
minimal spanning tree (MST) 339
minimum 73
Minkowski metric 229
mode 74

Mongo database system 307-313
MongoDB
    about 382-384
    download link 307
    indexing 328, 329
    need for 329
    references 308
MongoDB extension
    for geospatial databases 327
MongoDB installation file
    download link 382
MongoDB Manual
    reference 314
mongo-java-driver JAR files
    download link 318
moving average
    about 56
    computing, in Microsoft Excel 59, 60
MovingAverage class
    test program 57, 58
moving average series 57
multilinear functions 171
multiple linear regression 171-174
multivariate distributions 87-89
multivariate probability distribution
        function 87
MySQL
    database schema, creating 371, 372
MySQL database
    accessing, from NetBeans 373-376

N

naive Bayes classification algorithm 203
Neoj4 330
NetBeans
    MySQL database, accessing from 373-376
Netflix prize 300, 301
normal distribution
    about 65
    example 66
normal equations 150
NoSQL
    need for 329
    versus SQL 306, 307
NoSQL databases 303

NoSQL database systems
  about 330
  reference 329
null hypothesis 105
null values 15

O

offset 299
online recommender systems 267
ordinary least squares (OLS) 175
outlier 165

P

PageRank algorithm 332-337
parser 27
parsing 28
partitioning around medoids (PAM) 256
percentile 74
POI open source API library
  download link 20
polynomial regression 165-171
population 76
primary key 110
probabilistic events
  independence 90, 91
probabilities
  facts 78
probability density function (PDF) 67
probability distribution
      function (PDF) 79-81
probability function 77
probability set function 78
Pythagorean theorem 228

Q

quality control department (QCD) 103
quartiles 74

R

random access files
  using 298-300
random experiment 77
random sample 76

random sampling 76
random variable 79
range 73
recommender system 267
red-black tree data structure 295
Redis 330
regression 143
regression analysis 143
regression coefficients
  computing 148-150
relation 109
relational database design
  about 112, 113
  batch processing 130-133
  database, creating 114-117
  database queries 123
  database views 134-137
  data, inserting into database 121-123
  Java Database Connectivity (JDBC) 125-127
  JDBC PreparedStatement, using 128
  SQL commands 117-121
  SQL data types 125
  subqueries 137-139
  table indexes 140, 141
relational database (RDB) 109-111
relational database system (RDBMS) 114
relational database tables 15
relation data model 109
residual 147
rows 110
running average 56

S

sample 65
sample correlation coefficient 147
sample space 77
sample variance 73
scalability 344, 346
scaling 332
scatter plot
  about 46
  generating 47
schema 109, 111
scientific method 2
sharding 332

shards 332
show collections command
  collections 312
sigmoid curve 216
similarity measure 269-271
simple average 73
simple recommender system 272-283
skewness 76
SN 1572 2
sorting 36-38
sparse matrix 294, 336
sparse matrix format 347
spectacular example 4, 5
SQL commands 117-121
SQL data types 125
SQL script 120
SQL (Structured Query Language)
  about 115
  versus NoSQL 306, 307
Standard Generalized Markup Language (SGML) 24
standard normal distribution 96-101
statement object 128
statistics
  about 73
  descriptive statistics 73
subquery 137-139
support vector machine algorithms 209-214

T

table indexes 140, 141
tables 46
taxicab metric 230
test datasets
  generating 30
test set 180
time series
  about 51, 52
  simulating 68-70
TimeSeries class
  test program 53-56

total variation 151, 154
training set 180
transition matrix 334
triangle inequality 227
tuples 109
two-tailed test 107
Type I error 92, 105
Type II error 92
type signature 14

U

unexplained variation 154
Universal Product Codes (UPCs) 306
user ratings
  implementing 290-293
utility matrix 268, 269

V

variables 14
variation statistics 151-154
vehicle identification number (VIN) 305
vertical scaling 332
virtual table 134
VisiCalc 7

W

weighted mean 74
Weka
  about 198
  download link 381
Weka implementation 245-247
Weka libraries 381
Weka Workbench
  reference 198
WordCount example 339-344

X

XML data 24